Introduction to Qualitative Methods for Differential Equations

Introduction to Qualitative Methods for Differential Equations provides an alternative approach to teaching and understanding differential equations. The basic methodology of the book is centered on finding reformulations of differential equations in such a manner that they become (partially, at least) problems in geometry. Through this approach, the book distills the critical aspects of the qualitative theory of differential equations and illustrates their application to a number of nontrivial problems.

Features

- Self-contained with suggestions for further reading
- Concise and approachable exposition with only minimal prerequisites
- Ideal for self-study
- Appropriate for undergraduate mathematicians, engineers and other quantitative science students

Ronald E. Mickens is Emeritus Professor at Clark Atlanta University, Atlanta, GA, and is Fellow of several professional organizations, including the American Physical Society. He has written or edited 17 books and published more than 350 peer-reviewed research articles.

Introduction to Qualitative Methods for Differential Equations

Ronald E. Mickens

CRC Press
Taylor & Francis Group
Boca Raton London New York

CRC Press is an imprint of the
Taylor & Francis Group, an **informa** business

A CHAPMAN & HALL BOOK

Designed cover image: Shutterstock

First edition published 2025
by CRC Press
2385 NW Executive Center Drive, Suite 320, Boca Raton FL 33431

and by CRC Press
4 Park Square, Milton Park, Abingdon, Oxon, OX14 4RN

CRC Press is an imprint of Taylor & Francis Group, LLC

© 2025 Ronald E. Mickens

Reasonable efforts have been made to publish reliable data and information, but the author and publisher cannot assume responsibility for the validity of all materials or the consequences of their use. The authors and publishers have attempted to trace the copyright holders of all material reproduced in this publication and apologize to copyright holders if permission to publish in this form has not been obtained. If any copyright material has not been acknowledged please write and let us know so we may rectify in any future reprint.

Except as permitted under U.S. Copyright Law, no part of this book may be reprinted, reproduced, transmitted, or utilized in any form by any electronic, mechanical, or other means, now known or hereafter invented, including photocopying, microfilming, and recording, or in any information storage or retrieval system, without written permission from the publishers.

For permission to photocopy or use material electronically from this work, access www.copyright.com or contact the Copyright Clearance Center, Inc. (CCC), 222 Rosewood Drive, Danvers, MA 01923, 978-750-8400. For works that are not available on CCC please contact mpkbookspermissions@tandf.co.uk

Trademark notice: Product or corporate names may be trademarks or registered trademarks and are used only for identification and explanation without intent to infringe.

ISBN: 978-1-032-71598-8 (hbk)
ISBN: 978-1-032-72758-5 (pbk)
ISBN: 978-1-003-42241-9 (ebk)

DOI: 10.1201/9781003422419

Typeset in LM Roman
by Deanta Global Publishing Services, Chennai, India

Publisher's note: This book has been prepared from camera-ready copy provided by the authors.

This book is dedicated to my family
Maria | my wife,
Leah | my daughter,
James | my son.
Their collective interest, achievements, creativity and presence in my life have been a major factor in the decision to write this book.

Contents

Preface, xv

Preliminaries   1

  0.1 PURPOSE OF BOOK   1
  0.2 EXPERIMENTS AND PHYSICAL MEASUREMENTS   2
  0.3 MATHEMATICS AND PHYSICAL MATHEMATICS   2
  0.4 DYNAMIC CONSISTENCY AND MATHEMATICAL MODELING   3
  0.5 PHYSICAL AND MATHEMATICAL EQUATIONS   4
  0.6 LOCAL BEHAVIOR OF FUNCTIONS   6
  PROBLEMS   6
  NOTES AND REFERENCES   7

CHAPTER 1 ■ What Is a Solution?   10

  1.1 INTRODUCTION   10
  1.2 RADIOACTIVE DECAY   10
  1.3 $X^2 + Y^2 = 1$ 11
  1.4 MICKENS–NEWTON LAW OF COOLING   13
  1.5 WHAT ARE THE SOLUTIONS TO $Y'' + Y' = 0$?   14
  1.6 APPROXIMATE SOLUTION TO THE HEAT PDE   16
  1.7 DISCUSSION   18
  PROBLEMS   20
  NOTES AND REFERENCES   20

viii ■ Contents

Chapter 2 ■ One-Dimensional Systems	22
2.1 INTRODUCTION	22
2.2 FIXED-POINTS	22
2.3 SIGN OF THE DERIVATIVE: ONE FIXED-POINT	23
2.4 TWO FIXED-POINTS	24
2.5 LINEAR STABILITY	25
2.6 APPLICATIONS	27
2.6.1 Radioactive Decay	27
2.6.2 Logistic Equation	29
2.6.3 Gompertz Model	30
2.6.4 Draining a Tank	31
2.6.5 $f(x, t)$ Depends on t	33
2.6.6 Spruce Budworm Population Model	35
2.7 DISCUSSION	38
PROBLEMS	39
NOTES AND REFERENCES	39

Chapter 3 ■ Two-Dimensional Dynamical Systems	41
3.1 INTRODUCTION	41
3.2 DEFINITIONS	42
3.2.1 2-Dim Dynamical System	42
3.2.2 Fixed-Points	43
3.2.3 Nullclines	43
3.2.4 First-Integral and Symmetry Transformations	45
3.3 GENERAL FEATURES OF TRAJECTORIES	45
3.4 CONSTRUCTING PHASE-PLANE DIAGRAMS	49
3.5 LINEAR STABILITY ANALYSIS	50
3.6 LOCAL BEHAVIOR OF NONLINEAR SYSTEMS	51

3.7 EXAMPLES	53
3.7.1 Harmonic Oscillator	53
Comments	55
3.7.2 Damped Harmonic Oscillator	56
3.7.3 Nonlinear Cubic Oscillator	57
3.7.4 Damped Cube-Root Oscillator	60
3.7.5 $\ddot{x}+(1+\dot{x})\,x=0$	62
3.7.6 Simple Predator–Prey Model	65
3.7.7 van der Pol Equation	67
3.7.8 SIR Model for Disease Spread	69
3.8 DISCUSSION	74
PROBLEMS	74
COMMENTS AND REFERENCES	75

Chapter 4 ■ Sturm–Liouville Problems   76

4.1 INTRODUCTION	76
4.1.1 Elimination of First-Derivative Term	77
4.1.2 Liouville–Green Transformation	78
4.2 THE VIBRATING STRING	80
4.2.1 Both Ends Fixed	81
4.2.2 One Fixed and One Free Ends	82
4.2.3 Both Ends Free	83
4.2.4 Summary	84
4.3 SEPARATION AND COMPARISON RESULTS	85
4.3.1 $y''(x)+f(x)\,y(x)=0$	87
4.4 STURM–LIOUVILLE PROBLEMS	90
4.4.1 Properties of the Eigenvalues and Eigenfunctions	91
4.4.2 Orthogonality of Eigenfunctions	91
4.4.3 Expansion of Functions	92
4.5 RELATED ISSUES	93

4.5.1 Reduction to Sturm–Liouville Form	93
4.5.2 Fourier Series	94
4.5.3 Special Functions	96
4.5.4 TISE: Sketches of Wavefunctions	97
PROBLEMS	100
COMMENTS AND REFERENCES	101

CHAPTER 5 ■ Partial Differential Equations 102

5.1 GENERAL COMMENTS	102
5.2 SYMMETRY-DERIVED PDES	104
5.2.1 Heat Conduction PDE	104
COMMENTS	105
5.2.2 Wave PDE	106
5.2.3 Discussion	108
COMMENTS	109
5.3 METHOD OF SEPARATION OF VARIABLES	110
5.3.1 Introduction	110
5.3.2 Definition of the Method of SOV	111
5.3.3 Examples	113
5.4 TRAVELING WAVES	123
5.4.1 Burgers' Equation	124
5.4.2 Korteweg de Vries Equation	125
5.4.3 Fisher's Equation	128
Comments	130
5.4.4 Heat PDE	131
PROBLEMS	134
NOTES AND REFERENCES	134

Contents ■ xi

CHAPTER 6 ■ Introduction to Bifurcations	137
6.1 INTRODUCTION	137
6.2 DEFINITION	141
6.2.1 Bifurcation	141
6.3 EXAMPLES OF ELEMENTARY BIFURCATIONS	142
6.3.1 Saddle-node Bifurcation	143
6.3.2 Transcritical Bifurcation	144
6.3.3 Supercritical Pitchfork Bifurcation	147
6.3.4 Subcritical Pitchfork Bifurcation	148
6.4 EXAMPLES FROM PHYSICS	150
6.4.1 Lasers	150
6.4.2 Statistical Mechanics and Neural Networks	151
6.5 HOPF-BIFURCATIONS	152
6.5.1 Introduction	152
6.5.2 Hopf-Bifurcation Theorem	153
6.5.3 Fixed-Points and Closed Integral Curves	154
6.5.4 Two Limit-Cycle Oscillators	154
6.6 RESUMÉ	159
PROBLEMS	159
COMMENTS AND REFERENCES	160

CHAPTER 7 ■ Applications	161
7.1 ESTIMATION OF $Y(0)$ FOR A BOUNDARY-VALUE PROBLEM	161
7.1.1 Properties of $y(z)$	162
7.1.2 Approximation to $y(z)$	165
7.1.3 Resume	167
7.2 THOMAS–FERMI EQUATION (TFE)	168
7.2.1 Exact Results	168
7.2.2 Approximate Solutions	172

	7.2.3 Discussion	173
7.3	TRAVELING-WAVE FRONT BEHAVIOR FOR A PDE HAVING SQUARE-ROOT DYNAMICS	175
	7.3.1 Variable Scaling	177
	7.3.2 Traveling Wave Solutions	178
	7.3.3 Traveling Wave Front Behavior	180
	7.3.4 Case I	182
	7.3.5 Case II	182
	7.3.6 Case III	183
	7.3.7 Approximation to Traveling Wave Solution	184
7.4	COMMENTS ON FUNCTIONAL EQUATION MODELS OF RADIOACTIVE DECAY AND HEAT CONDUCTION	185
7.5	APPROXIMATE SOLUTIONS TO A MODIFIED, NONLINEAR MAXWELL–CATTANE EQUATION	194
	7.5.1 Positivity and Equilibrium Solutions	196
	7.5.2 Space-Independent Solutions	197
	7.5.3 Traveling Waves	200
	7.5.4 Resume	205
7.6	NONLINEAR OSCILLATIONS: AN AVERAGING METHOD	206
	7.6.1 First Approximation of Krylov and Bogoliubov	206
	7.6.2 Higher-Order Corrections	209
	7.6.3 Examples	209
	7.6.4 Summary	222
7.7	CULLING IN PREDATOR–PREY SYSTEMS	222
	7.7.1 Predator–Prey Models	223
	7.7.2 General Properties of Predator–Prey Models	224
	7.7.3 Culling	227
	7.7.4 Culling the Predator	227
	7.7.5 Summary	228
7.8	A LINEAR ODE: $Y' = (X - Y)/X^2$	229

	7.8.1 Qualitative Analysis	231
	7.8.2 Construction of an Approximate Solution	233
	7.8.3 Summary	235
7.9	APPROXIMATING '1' AND '0'	236
	7.9.1 Introduction	236
	7.9.2 Finite Difference Discretization of a Modified Decay ODE	237
	7.9.3 $d^2x/dt^2 + x^3 = 0$	240
	7.9.4 $\ddot{x} + x^{\frac{1}{3}} = 0$	243
	7.9.5 Discussion	245
	COMMENTS AND REFERENCES	246
	REFERENCES TO THE EXPONENTIAL FUNCTIONS	250

Appendix A 253

A.1	ALGEBRAIC RELATIONS	253
	A.1.1 Factors and Expansions	253
	A.1.2 Quadratic Equations	253
	A.1.3 Cubic Equations	254
	A.1.4 Expansions of Selected Functions	255
A.2	TRIGONOMETRIC RELATIONS	255
	A.2.1 Fundamental Properties	255
	A.2.2 Sum of Angles Relations	256
	A.2.3 Product and Sum Relations	256
	A.2.4 Derivatives and Integrals	257
	A.2.5 Powers of Trigonometric Functions	257
A.3	HYPERBOLIC FUNCTIONS	257
	A.3.1 Definitions and Basic Properties	257
	A.3.2 Derivatives and Integrals	257
	A.3.3 Other Important Relations	258
	A.3.4 Relations between Trigonometric and Hyperbolic Functions	258

A.4	IMPORTANT CALCULUS RELATIONSHIPS	258
	A.4.1 Differentiation	258
	A.4.2 Integration by Parts	259
	A.4.3 Differentiation of a Definite Integral	259
A.5	EVEN AND ODD FUNCTIONS	259
A.6	ABSOLUTE VALUE FUNCTION	260
A.7	DIFFERENTIAL EQUATIONS	261
	A.7.1 General Linear, First-Order Ordinary Differential Equation	261
	A.7.2 Bernoulli Equations	261
	A.7.3 Riccati Equation	262
	A.7.4 Linear, Second-Order Differential Equations with Constant Coefficients	262
	A.7.5 Fourier Series	264

BIBLIOGRAPHY, 266

INDEX, 268

Preface

My main purpose in writing this book was to introduce some members of the scientific community to a set of mathematical tools that can be used to obtain qualitative information about the solutions of differential equations. These techniques are important because they allow the determination of many features of the solutions independent of a knowledge of the exact solutions. With this general overview of the solutions and their critical features, generally 'good mathematical approximations' can be constructed and applied to analyze the physical system modeled by the original differential equations.

This book consists of eight short chapters and an appendix of useful mathematical relations. The major issues discussed include the following topics:

What is a solution to a differential equation?

Linear stability of 1-dim and 2-dim systems

Sturm–Liouville problems and the Liouville–Green transformation

Separation and comparison theorems for differential equations

Separation of variable and traveling wave solutions for partial differential equations

Introduction to bifurcation techniques

Limit cycles

Approximating the numbers '0' and '1'

How many of you know that the numbers, '0' and '1', can be expanded as a parameter, 'p', and used (via perturbation techniques) to calculate approximate solutions to differential equations?

To make good use of this book, the reader needs only to have the knowledge that comes from having an undergraduate degree in one of the physical, mathematical and/or engineering sciences. Of course, having a background at higher levels of technical education will help.

The general perspective of this book is that of a scientist, not that of a mathematician concerned with rigorous proofs. In general, no proofs of anything are given. However, when required, relevant references are provided.

As I have stated in several previous books, all published by Taylor & Francis/CRC Press, I thank Callum Fraser, my Editor, for allowing me complete freedom to write this book in the unconventional style and format that you are accessing.

Finally, I want to acknowledge two others who have provided their services to get me a large number of articles and books related to the subject matter of this book. They are Ms Imani Beverly and Mr Brian Briones, both reference librarians at the Atlanta University Center's Robert W. Woodruff Library, Atlanta, Georgia. Without their effective and speedy aid, the completion of this book would have been greatly delayed.

Ronald E. Mickens
Atlanta, Georgia 30314 USA

Preliminaries

0.1 PURPOSE OF BOOK

Humans are curious, especially concerning the physical universe. Science began when (some) humans understood that ignorance exists, but they could do something about it within the context of a process that we now call the scientific methodology. Jointly with this methodology there arose an associated language, which we now call 'mathematics'. One consequence of the construction of mathematical tools was their separate evolution into an intellectual field for the study of its own features, separate from its origins in science.

Taking mathematics as the language of science, the precision study of the physical universe began with the construction of mathematical models for the analysis of particular physical phenomena. However, an interesting thing happened; essentially none of the associated mathematical equations could be solved exactly for their solutions. Consequently, a broad range of approximation techniques were created including computational methods for determining numerical solutions.

Another, less used set of techniques, is based on acquiring only qualitative information on the desired solutions. While many of these procedures have been available since the early nineteenth century, many working scientists and researchers are not knowledgeable about their existence and/or how to apply them to the equations they hope to analyze, understand and solve.

The main purpose of this small volume is to introduce some of these qualitative methods and show how they may be applied to differential equations. The reason for the focus on differential equations follows from the fact that many, if not most, of the physical systems have some form of mathematical model for which the mathematical structure is represented by differential equations.

0.2 EXPERIMENTS AND PHYSICAL MEASUREMENTS

Experimental science and the data it generates by means of physical measurement have an important aspect: it is discrete in nature. All data acquisition, often in some hidden manner, relies on making space and time measurements and these are always done discretely.

0.3 MATHEMATICS AND PHYSICAL MATHEMATICS

There is no standard agreement as to how to define mathematics. As a practical matter, I use the following working definition

Mathematics is the study, creation and analysis of patterns in the abstract universe of human thought and mental perception.

Likewise, my working definition of science is

> Science is the systematic observation, creation, analysis, and modeling of patterns which exist in the physical universe.

However, Albert Einstein's statement on the relationship between science and mathematics must be taken very seriously. He stated that

> as far as the laws of mathematics refer to reality, they are not certain; and as far as they are certain, they do not refer to reality.

The definitions of mathematics and science, given above, along with Einstein's statement, imply that current mathematical structures are not the 'exact language' of science, as many believe. While mathematics and science are deeply intertwined, neither is an exact replica of the other. At best, current mathematics is an approximation to whatever may be the 'natural language of science'. What can be said is that mathematics in many instances works extremely well for formulating and expressing science and its results, but what we have now available is not the final word. What is required is a new mathematical structure or formulation that directly incorporates how experiments are actually done and data acquired. Further, not only are all experiments discrete in nature, they also have errors occurring as part of the measuring processes.

Physical mathematics is the application to the analysis of physical systems' concepts, ideas, techniques and understandings that arise in mathematics, but without the strict logical vigor that we associate with mathematics. For example, for physical mathematics

There are no real numbers, only rationals...

Both independent and dependent variables are discrete valued...

Physical infinities are assumed not to exist...

There do not exist derivatives; differences are what is measured....

These 'facts' imply that calculus and its generalizations to differential equations, Fourier theory, complex function theory, geometry, etc., are approximations. The required mathematical structure needed to faithfully represent the physical universe has to be some discrete form of these 'continuous structures'. And, for the time being, we have no clue as to what they are!

In recent years, the idea that nature is discrete or that nature can only be realistically analyzed from the perspective of a discrete-based mathematical structure has been presented to the scientific community. Essays giving arguments, both mathematical and physical, against nature being continuous have been given by Gregory Chaitin and Sheldon Glashow.

0.4 DYNAMIC CONSISTENCY AND MATHEMATICAL MODELING

We now provide some definitions related to just what is mathematical modeling and the role played by dynamic consistency in the formulation of these models. Our discussion here is brief and the reader is referred to Mickens book (2022) for a more detailed presentation.

Definition 1 A *system* is a set of interrelated, interacting, interdependent elements, which collectively form a complex whole.

Definition 2 *Modeling* is the representation of one system by another system.

Definition 3 *Mathematical modeling* is the representation of a system by a set of mathematical relations or equations.

Definition 4 *Dynamic consistency* is concerned with the relationship between two systems S_1 and S_2. Let S_1 have the property or feature P. If S_2 also has property P, then S_2 is said to be dynamically consistent with S_1 with respect to P.

Examples of properties for which two systems might be dynamically consistent include the following items:

Satisfaction of a conservation law

Positivity of particle number or density

Stability properties of equilibrium states

etc.

For our purposes S_1 could be a physical system, s_2 a mathematical model and S_3 the solutions of the equations derived to obtain S_2, i.e., in pictorial representation.

S_1: physical system
$\downarrow$
S_2: mathematical model
$\downarrow$
S_3 : Solutions

Note that S_2 may be dynamically consistent with respect to certain features of S_1, but not to other features. It is this fact that allows the creation of a range of mathematical models for S_1. Similarly, S_3 may be dynamically consistent with particular features of S_2, but not with other aspects. This realization is the basis of many of the issues related to determining analytic approximation to the solution of mathematical equations and the introduction of numerical errors in discretization schemes.

Consequently, given a system S, different models of it, $\overline{S}$ can be obtained by creating/constructing models $(\overline{S}_1, \overline{S}_2, ...)$, which are dynamically consistent with different combinations of features or properties of S. This fact is an illustration of the fact that the modeling process is generally not unique.

BTW: In this book, essentially all the mathematical models are represented as differential equations.

0.5 PHYSICAL AND MATHEMATICAL EQUATIONS

Usually the mathematical modeling equations are derived from various physical principles. This means that their independent and dependent variables and associated parameters are expressible in terms of the physical units of (mass, length, time). However, it is generally more suitable to transform these equations into expressions where all the variables and parameters have no physical units. The main reason for doing this is that the physical variables and parameter magnitudes will be dependent on the units of physical measurement selected. To resolve this issue, the notion of scaled variables is introduced.

Definition 5 A *physical equation*, derived in the construction of a mathematical model for a system, in general, has all its variables and parameters expressed in terms of a given set of physical units such as mass, length and time.

Definition 6 A *mathematical equation* has all of its variables and parameters dimensionless, i.e., they can be expressible as pure numbers.

The major advantages of a mathematical equation are that, in general, they contain fewer parameters and this makes the resulting equations easier to study. The following elementary example illustrates this point.

Consider the decay equation

$$\frac{dx}{dt} = -\lambda x, x(0) = x_0 \text{ (given)}, \qquad (0.5.1)$$

where λ is a positive parameter. If x has units of length and t, units of time, then λ has units of inverse time, i.e.,

$$[x] = L, [t] = T, [\lambda] = \frac{1}{T}. \qquad (0.5.2)$$

The original differential equation can be rewritten to the form

$$\frac{dy(\bar{t})}{d\bar{t}} = -y(\bar{t}), y(0) = 1, \qquad (0.5.3)$$

where

$$t \to \bar{t} = \lambda t, x(t) \Rightarrow y(\bar{t}) = \frac{x(t)}{x_0}, \qquad (0.5.4)$$

Note two things about Equation (0.5.3). First, its variables are dimensionless or pure numbers. Second, for this example, the y-equation has only one, simple initial condition, $y(0) = 1$. This certainly aids in the overall analysis of the solution behavior of the decay equation. This type of manipulation can be extended to more complex equations.

0.6 LOCAL BEHAVIOR OF FUNCTIONS

Much of this book will be, at some point, based on the ability to sketch representations of various functions *by hand*. To effectively do so, we will need to know some basic features of the properties of functions. One of the things to aid us in this task is to understand the local behaviors of functions, $f(x)$, near a point, $x = x_1$, based on the properties of its derivative or slope. Fortunately, nearly all the functions that we will consider are at least piece-wise-continuous.

Let

$$y(x) = f(x), \qquad (0.6.1)$$

and consider its plot, i.e., $y(x)$ vs x, in the $x - y$ plane. Assume that at $x = x_1$, the derivative, $y'(x) = df(x)/dx$, exists. Figure 0.1 gives the local behavior of $y = f(x)$. For various values of $y'(x_1)$. For completeness, we have also included the two cases where the slope is unbounded or infinite. In words, Figure 0.1 depicts the following situations:

(1) If at $y_1 = y(x_1)$, $y'(x_1) > 0$, then the curve passing through the point, (x_1, y_1), has a tangent that is pointing upward and to the right.

(2) If $y'(x_1) = 0$, then the tangent is horizontal and pointing to the right.

(3) If $y'(x_1) < 0$, then the tangent is pointing downward and to the right.

(4) If $y'(x_1) = +\infty$, then the tangent is pointing vertically upward.

(5) If $y'(x_1) = -\infty$, then the tangent is pointing vertically downward.

PROBLEMS

Section 0.2

1) Think about how the number 'π' is used in mathematics and computations. Is the 'same' numerical π appearing in each situation?

Section 0.4

2) Discuss why physical mathematics and mathematics are not equivalent.

3) It is often stated that 'Mathematics is the language of science'. What does this statement mean? Is it true?

Section 0.5

4) The van der Pol oscillator differential equation is

$$m\frac{d^2x}{dt^2} + kx = (a - bx^2)\frac{dx}{dt},$$

where m has the dimension of mass, x and t have, respectively, the dimensions of length and time. Scale this ODE such that all the new variables and parameters have no physical dimensions.

Section 0.6

5) Consider the curve

$$x^2 + y^2 = 1.$$

Locate and draw this curve indicating points where

$$\dot{y} = 0, \dot{y} = \infty, \dot{y} = -\infty, \dot{y} = 1, \dot{y} = -1.$$

NOTES AND REFERENCES

Some references that cover in more detail the variety of topics discussed in this chapter are listed below:

1. R. E. Mickens, Dynamic consistency: A fundamental principle for constructing NSFD schemes for differential equations, *Journal of Difference Equations and Applications*, Vol. 11 (2005), 645–653.

2. R. E. Mickens, *Mathematical Methods for the Natural and Engineering Sciences*, 2nd Edition (World Scientific, London, 2017).

3. T. Szirtes, *Applied Dimensional Analysis and Modelling* (MaGraw-Hill, New York, 1998).

8 ■ Introduction to Qualitative Methods for Differential Equations

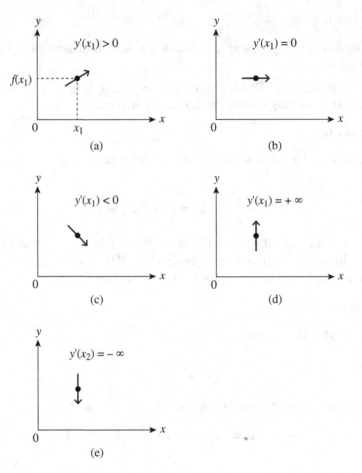

FIGURE 0.1 Local behavior of the curve, $y = f(x)$, in the $x - y$ plane for $x = x_1$. For, all are drawn for the first

Arguments and discussions on the nature of science and mathematics, and their relations to each other are covered in the following interesting references:

4. I. Bah, D. S. Freed, G. W. Moore, N. Nekrasov, S. S. Razamat, and S. Schafer-Nameki, A panorama of physical mathematics; published in many venues, but also arXiv:2211.04467v3 [hep-th] (Accessed May 9, 2024).

5. G. Chaitin, How real are real numbers?; arXiv:math/0411418v3 [math.HO] (Accessed November 29, 2004).

6. G. t'Hooft, Confusions regarding quantum mechanics, reply by sheldon Lee Glashow, *Inference*, Vol. 5, No. 3 (September 2020). https://doi.org/10.37282/991819.20.56.

7. B. Wu, Mathematics is physics; arXiv:2306.03766v2 [physics.gen-ph] (Accessed June 7, 2023).

CHAPTER 1

What Is a Solution?

1.1 INTRODUCTION

This chapter examines the process of 'calculating' solutions to the differential equations modeling physical phenomena. The various obtained results will not necessarily be what a mathematician would either calculate or find rigorous. Our point of view is that of a scientist whose task is to find mathematics-based structures that provide physical insight into the phenomena under investigation. While the modeling equations are relatively elementary, they illustrate the methodology of a mathematical scientist who is doing science rather than being a mathematician. Remember, insights and understanding are the goals.

1.2 RADIOACTIVE DECAY

Radioactive decay is a general phenomenon occurring in many areas of the natural and engineering sciences. For elementary decay, it is the usual case that it is modeled by a first-order differential equation having the form

$$\frac{dx}{dt} = -\lambda x, \quad x(0) = x_0, \qquad (1.2.1)$$

where λ has the physical units of (time)$^{-1}$ and $x(t)$ is the amount of radioactive decaying material at time t.

Another approach, which is more general, is to construct a mathematical model based on the experimental finding that for elementary decays, sometimes called 'direct decays', there exists a constant τ such that at time $t + \tau$, the amount of radioactive material is one-half the amount at time t. Mathematically, we can express this as the following relationship

$$x(t+\tau) = \left(\frac{1}{2}\right)x(t), \quad x(0) = x_0. \tag{1.2.2}$$

If we denote the respective solutions of Equations (1.2.1) and (1.2.2) by $x_1(t)$ and $x_2(t)$, then as shown by Mickens

$$x_1(t) = x_0 e^{-\lambda t}, \tag{1.2.3}$$

$$x_2(t) = x_0 e^{A(t)} e^{-\left[\frac{Ln(2)}{\tau}\right]t}, \tag{1.2.4}$$

where

$$A(-t) = -A(t), \quad A(t+\tau) = A(\tau). \tag{1.2.5}$$

Note that $A(t)$ is an odd function and periodic with period τ. The simplest case is $A(t) = 0$, giving Equation (1.2.3) with

$$\lambda = \frac{Ln(2)}{\tau}. \tag{1.2.6}$$

Inspection of Equation (1.2.2) shows that it is a functional equation, in contrast to Equation (1.2.1), and has a much broader set of solutions. The main reason why the differential equation model is used to study simple decay is the fact that it has a solution that is simple to comprehend and the exponential form requires only one parameter, λ, to uniquely determine its behavior. On the other hand, the functional Equation (1.2.2) actually requires prior knowledge of $X(t)$ over a time interval τ to uniquely determine its solutions, which is essentially impossible to do by means of experiments. This clearly demonstrates that what needs to be done is superseded by what can actually be done easily.

1.3 $X^2 + Y^2 = 1$

What does this equation represent? Another way of stating this question is 'What does this expression mean?'
One interpretation of

$$x^2 + y^2 = 1 \tag{1.3.1}$$

12 ■ Introduction to Qualitative Methods for Differential Equations

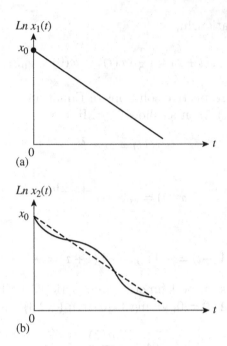

FIGURE 1.1 Plots of the decay curves: (a) is for Equation (1.2.3). (b) is for Equation (1.2.4), where the oscillations are exaggerated.

is that it represents the plot of a unit circle in the x–y plane.

A second interpretation is that it represents two functions $f(t)$ and $g(t)$ such that the relationship between them is given by Equation (1.3.1). Particular examples include

$$x = f_1(t) = \cos t, \quad y = g_1(t) = \sin t, \tag{1.3.2}$$

and

$$x = f_2(t) = CNt, \quad y = g_2(t) = SNt, \tag{1.3.3}$$

where $f_2(t)$ and $g_2(t)$ are, respectively, the Jacobian cosine and sine functions. In fact, an unlimited number of periodic functions can be constructed such that they satisfy Equation (1.3.1); for example

$$x(t) = \cos\theta(t), \quad \sin(t) = \sin\theta(t), \tag{1.3.4}$$

$$\theta(-t) = -\theta(t), \quad \theta(t+T) = \theta(t) + 2\pi, \tag{1.3.5}$$

where
$$\theta(t) = 2\pi\left(\frac{t}{T}\right) + A(t), \qquad (1.3.6)$$
with
$$A(-t) = -A(t), \qquad A(t+T) = A(t). \qquad (1.3.7)$$
The period of these functions is T.

Assuming x and y to be functions of t and taking the derivative of 1.3.1, we obtain
$$x\frac{dx}{dt} + y\frac{dy}{dt} = 0. \qquad (1.3.8)$$
From this, it follows that
$$\frac{dx}{dt} = h(t)\,y, \qquad \frac{dy}{dt} = -h(t)\,x, \qquad (1.3.9)$$
where we assume for convenience that
$$h(t) > 0 \text{ for all relevant } t. \qquad (1.3.10)$$
Thus, given any such $h(t)$, $x(t)$ and $y(t)$ are periodic. The simplest case is for $h(t) = 1$, giving
$$\frac{dx}{dt} = y, \qquad \frac{dy}{dt} = -x, \qquad (1.3.11)$$
whose solutions are
$$x(t) = \sin t, \qquad y(t) = \cos t. \qquad (1.3.12)$$
In summary, the answer to the question 'What is the solution to $x^2 + y^2 = 1$?' depends on the context in which the question is asked?

1.4 MICKENS–NEWTON LAW OF COOLING

Newton's law of cooling is not dynamically consistent with the fact that objects that are cooling or heating do not take an unlimited time to reach the equilibrium temperature of their environments. Mickens has created a modified model to deal with this situation. In its simplest manifestation, it takes the form
$$\frac{dx}{dt} = -\lambda x^{\frac{1}{3}}, \qquad x(0) = x_0, \qquad (1.4.1)$$

14 ■ Introduction to Qualitative Methods for Differential Equations

where

$$T_e = \text{fixed environmental temperature,}$$
$$T_0 = \text{initial temperature of the object,}$$
$$T(t) = \text{temperature of the object at time} t,$$
$$x(t) = T(t) - T_e,$$

Note that

$$\lim_{t \to \infty} T(t) = T(\infty) = T_e, \qquad (1.4.2)$$

Also, Equation (1.4.1) is a separable, first-order, nonlinear differential equation and can be easily integrated to give the solution

$$x(t) = \begin{cases} \left[x_0^{\frac{2}{3}} - \frac{2\lambda t}{3} \right]^{\frac{3}{2}}, & 0 < t \le t^*, \\ 0, & t > t^*, \end{cases} \qquad (1.4.3)$$

where

$$t^* = \frac{3 x_0^{\frac{2}{3}}}{2\lambda}. \qquad (1.4.4)$$

I have had several really outstanding mathematicians tell me that Equation (1.4.1) does not satisfy existence–uniqueness conditions, and therefore this equation cannot provide a valid physical model! Can you think of reasons why their analysis is incorrect? In any case, it requires a deep understanding of the rather obvious physics to go from the mathematical modeling equation, Equation (1.4.1), to the solution given by Equations (1.4.3) and (1.4.4). All of this is connected to the fact that $x(t) = 0$ is an equilibration solution.

1.5 WHAT ARE THE SOLUTIONS TO $Y'' + Y' = 0$?

This differential equation can to used to model a number of physical systems, including a massive object moving in a viscous liquid and having no other force acting on it. Since it is a second-order, linear differential equation, the initial conditions will be selected to be

$$y(0) = y_0 > 0, \quad y'(0) = -y_0' < 0, \qquad (1.5.1)$$

where $y = y(t)$ and we write out in full our differential equation

$$\frac{d^2y}{dt^2} + \frac{dy}{dt} = 0. \qquad (1.5.2)$$

Assuming that the solutions take the form

$$y(t) = e^{rt}, \qquad (1.5.3)$$

we obtain the characteristic equation

$$r^2 + r = r(r+1) = 0. \qquad (1.5.4)$$

Therefore, two particular solutions are

$$y_1(t) = 1, \quad y_2(t) = e^{-t}, \qquad (1.5.5)$$

Since the solutions to Equation (1.5.4) are

$$r_1 = 0, \quad r_2 = -1. \qquad (1.5.6)$$

With these results, the general solution to Equation (1.5.2) is

$$y(t) = C_1 + C_2 e^{-t}, \qquad (1.5.7)$$

where C_1 and C_2 are arbitrary constants.

It should be indicated that the general solution $y(t)$, as given by Equation (1.5.7), has derivatives of all orders for $-\infty < t < +\infty$, and this may be acknowledged by stating that the solution to Equation (1.5.2) is a $C^{(\infty)}$ function. However, since the original differential equation is only second order, we really only need to require its solutions to be $C^{(1)}$, i.e., the function $y(t)$ and its derivative, dy/dt, exist and are continuous. However, if we *allow* the existence of piecewise *continuous functions*, then solutions exhibited in Figure 1.2 also can occur. Some of these solutions are $C^{(0)}$ and are not special cases of the general solution, Equation (1.5.7), and we will call these solutions generalized solutions of the original differential equation. However, most mathematicians would take the view that this manner of classifying 'solutions' is nonsense and clearly not mathematically valid. But, our view is that such functions still might be physically useful in interpreting the behavior of the universe. Remember that we are doing science rather than mathematics and in science one tries everything.

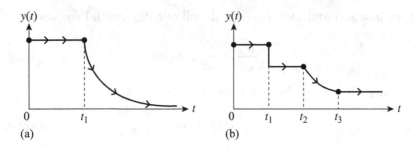

FIGURE 1.2 Two piecewise continuous or generalized solutions of Equation (1.5.2).

1.6 APPROXIMATE SOLUTION TO THE HEAT PDE

The heat equation, in space dimension, is

$$u_t = Du_{xx}, \quad u(x,0) = f(x) \tag{1.6.1}$$

where D is the assumed constant diffusion coefficient and the following initial conditions hold

$$u(0,t) = 0, \quad u(L,t) = 0. \tag{1.6.2}$$

Physically, these equations model a wire of length L, that is insulated, and the temperature (u) of the ends, $x = 0$ and $x = L$, is maintained at the value zero.

Rescaling the independent variables

$$x \to L^*\bar{x}, \quad t \to T^*\bar{t}, \tag{1.6.3}$$

where L^* and T^* are scaling factors, gives

$$\frac{\partial u}{\partial \bar{t}} = \frac{\partial^2 u}{\partial \bar{x}^2}, \quad 0 \le \bar{x}^b \le 1. \tag{1.6.4}$$

For

$$L^* = L, \quad T^* = \frac{L^2}{D}. \tag{1.6.5}$$

For the remainder of this section, we will drop all bars on the variables and use the equation

$$\frac{\partial u}{\partial t} = \frac{\partial^2 u}{\partial x^2}; \quad u(0,t) = 0, u(1,t) = 0, u(x,0) = f(x), \tag{1.6.6}$$

where $f(x)$ is specified with

$$f(0) = 0, \quad f(1) = 0; \quad f(x) > 0, \quad 0 < x < 1. \tag{1.6.7}$$

Let us assume that we have no knowledge as to how Equation (1.6.6) can be solved. So, what to do? One rather elementary possibility is to assume that $u(x,t)$ takes the form of a product of two functions, one depending only on x and the other only on t, i.e.,

$$u(x,t) = F(x)\,G(t). \tag{1.6.8}$$

Substitution of this ansatz into the PDE in Equation (1.6.6) gives

$$F(x)\,G'(t) = F''(x)\,G(t). \tag{1.6.9}$$

Now, let us try something new or at least different. Assume that $F(x)$ is known and average of both sides of the equation over x, i.e.,

$$\left[\int_0^1 F(x)\,dx\right] G'(t) = \left[\int_0^1 F''(x)\,dx\right] G(t). \tag{1.6.10}$$

If we define the constants (a, b) by the relations

$$a = \int_0^1 F(x)\,dx, \quad b = \int_0^1 F''(x)\,dx, \tag{1.6.11}$$

then it follows that $G(t)$ satisfies the first-order differential equation

$$\frac{dG}{dt} = \left(\frac{b}{a}\right) G. \tag{1.6.12}$$

At this point, the essential issue is how to select the function $F(x)$ from which the constants (a, b) can be calculated. Note that $F(x)$ must be such that

$$\frac{b}{a} < 0. \tag{1.6.13}$$

But, where does this condition come from? It follows from the experiences that we have had with hot and cold bodies that generally a hot body decreases in time its temperature (u = temperature). Consequently, $dG/dt < 0$.

With nothing else to go on, select

$$F(x) = f(x) = u(x, 0), \tag{1.6.14}$$

where
$$f(x) = x(1-x^2). \tag{1.6.15}$$

Therefore,
$$a = \int_0^1 x(1-x^2)\,dx = \frac{1}{4}, \tag{1.6.16}$$

$$b = \int_0^1 (-6x)\,dx = -6 \tag{1.6.17}$$

and
$$\frac{dG}{dt} = -24G \Longrightarrow G(t) = Ae^{-24t}, \tag{1.6.18}$$

where A is an arbitrary integration constant. Hence
$$\begin{aligned}U(x,t) &= F(x)\,G(t) \\ &= Ax(1-x^2)\,e^{-24t}, \end{aligned} \tag{1.6.19}$$

Since $U(x,0) = x(1-x^2)$, we have $A = 1$ and the approximation to the solution of Equation (1.6.6) is
$$u(x,t) = x(1-x^2)\,e^{-24t}. \tag{1.6.20}$$

In summary, while it should not be expected that this approximation to the solution of Equation (1.6.6) is exact, it still might be reasonable to hold the view that this formula, i.e., Equation (1.6.20), at least provides good qualitative insight into both the physics represented by this model and the exact solution to Equation (1.6.7).

Later, we will consider the nonlinear oscillations of a beam and see how its time dependence behavior can be well approximated by the application of the methodology presented in this section.

1.7 DISCUSSION

We have not really answered the issue of 'What is a solution to a differential equation?' For a partial differential equation, whether linear or nonlinear, there is no concept of a general solution and this situation can generate confusion as to what mathematical expression to use for a given application. However, things are somewhat better for ordinary differential equations; but even for them there may exist singular solutions that are not special cases of the so-called general solutions.

This chapter examined a number of toy model differential equations, which appear in the modeling of some important physical systems. Our discussions hint at the fact that the mathematical models may not be unique: an example is that for decay in radioactivity. For such cases, the interpretation of the solutions coming from the different models implies that what experiments are done have to be carefully selected to provide useful data to actually distinguish between the mathematical models.

The solution technique presented in Section 1.6 illustrates both the power and weakness of ad hoc methods to obtain either approximate or exact solutions to an arbitrary differential equation. For our particular case, we examined the heat equation and made a definite assumption regarding the mathematical structure of a particular solution. However, it should be clearly understood that the selected ansatz depends on our needs and knowledge. What is selected is neither 'right' or 'wrong'. Its value derives from whatever insights it can provide about the physical system being investigated.

Finally, we should take very seriously the following comments made by Edward Redish:

> Mathematics is commonly referred to as 'the language of science' ... But, using math in science ... is not just doing math. It has a different purpose – representing meaning about physical systems rather than expressing abstract relations – and it even has a distinct semiotics - the way meaning is put into symbols – from pure mathematics.
>
> It almost seems that the 'language' of mathematics we should in physics is not the same as the one taught by mathematicians.

PROBLEMS

Sections 1.2 and 1.4

1) Consider the following modified decay and/or coding equation

$$\frac{dx}{dt} = -\lambda x - \epsilon x^{\frac{1}{3}},$$

(λ, ϵ) are positive parameters.
Solve this first-order, nonlinear differential equation and analyze the properties of its solutions.

Section 1.5

2) Let the solution to a linear ordinary differential equation be

$$y(t) = c_1 + c_2 e^{-t} + c_3 e^{-5t},$$

(C_1, C_2, C_3) are arbitrary constants. Construct the differential equation which has this solution. Sketch some bounded, continuous, piecewise functions on the interval, $-\infty < t < +\infty$.

Section 1.6

3) Consider the nonlinear diffusion PDE

$$uu_t = Du_{xx},$$

$D > 0$, is a constant. Construct solutions having the form $U(x,t) = f(x)g(t)$.

NOTES AND REFERENCES

Section 1.2: Modeling of radioactive decay is discussed in the article.

(1) R. E. Mickens and S. Rucker, A note on a functional equation model of decay processes: Analysis and consequences, *Journal of Difference Equations and Applications* (October 5, 2023). https://doi.org/10.1080/10236198.2023.2260891.

Section 1.3: The methodology of determining some of the solutions to $x^2 + y^2 = 1$ is presented in the following book:

(2) R. E. Mickens, *Generalized Trigonometric and Hyperbolic Functions* (CRC Press, Boca Raton, FL, 2019).

Section 1.4: For a discussion of various cooling laws and their solutions (which can also apply to the mathematics given in Section 1.2) see

(3) U. Besson, The history of the cooling law: When the search for simplicity can be an obstacle, *Science and Education*, Vol. 21, No. 8 (2010), 1−26.

(4) W. Dule and R. E. Mickens, Exact and Nonstandard finite difference schemes for a modified law of cooling, *Georgia Journal of Science*, Vol. 77, No. 1, Article 120. https://digitalcommons.gaacademy.org/gjs/vol77/iss1/120.

(5) R. E. Mickens and K. Oyedeji, Investigation of pawer-law damping/dissipative forces; arXiv:1405.4062v1 [physics. comp-ph] May 16, 2014.

Section 1.6: My search of the literature turned up essentially no wide use of the averaging technique presented in this section. However, for the field of nonlinear oscillations, see

(6) I. Kovacic and M. J. Brennan, *The Duffing Equation: Nonlinear Oscillators and Their Behaviors* (Wiley, New York, 2011). See Chapter 1.

Section 1.7: The following references are of relevance to some of the issues discussed in this section

(7) A. D. Polyanin and V. F. Zaitsev, *Handbook of Exact Solutions for Ordinary Differential Equations* (CRC Press, Baco Rotan, FL, 2002).

(8) S. V. Meleshko, *Methods for Constructing Exact Solutions of Partial Differential Equations* (Springer, New York, 2005).

(9) E. F. Redish, Problem solving and the use of math in physics courses; arXiv:physics/0608268v1[physics.ed-ph]

CHAPTER 2

One-Dimensional Systems

2.1 INTRODUCTION

If the mathematical model of a continuous dynamical system is represented by a single, first-order differential equation

$$\frac{dx}{dt} = f(x), \quad x(0) = x_0 \text{given}, \qquad (2.1.1)$$

then it will be called a one-dimensional dynamic system. Note that $f(x)$ depends only on x and not the independent variable t. This type of differential equation is called a first-order, autonomous equation. We also assume that $f(x)$ has mathematical properties such that the sought-after physical solutions exist and are unique.

2.2 FIXED-POINTS

The fixed-points of Equation (2.1.1) correspond to its constant solutions, i.e.,

$$x(t) = \bar{x} = \text{constant}. \qquad (2.2.1)$$

Consequently, they are solutions of the equation

$$f(x) = 0. \qquad (2.2.2)$$

In general for physical systems, only the real zeros have physical meaning. Further, these fixed-points correspond to the equilibrium or time-independent states of the system.

2.3 SIGN OF THE DERIVATIVE: ONE FIXED-POINT

Consider the case where there is a single, simple fixed-point at $x = \bar{x}$, i.e., Equation (2.1.1) takes the form

$$f(x) = f_1(x)(x - \bar{x}). \tag{2.3.1}$$

This implies that $f_1(x)$ has a definite sign for all physically relevant x, i.e., either $f_1(x) > 0$ or $f_1(x) < 0$.

Case A: $f_1(x) > 0$
We have

$$\begin{cases} \dfrac{dx}{dt} > 0, \text{if } x > \bar{x}, \\ \dfrac{dx}{dt} < 0, \text{if } x < \bar{x}. \end{cases} \tag{2.3.2}$$

Case B: $f_1(x) < 0$
Likewise, we have

$$\begin{cases} \dfrac{dx}{dt} < 0, \text{if } x > \bar{x}, \\ \dfrac{dx}{dt} > 0, \text{if } x < \bar{x}. \end{cases} \tag{2.3.3}$$

Following Equations (2.3.2) and (2.3.3), we have depicted the corresponding motions of the system along the x-axis. Note that $dx/dt > 0$ means moving to the right, while $dx/dt < 0$ implies motion to the left.

For a double zero, we have

$$\frac{dx}{dt} = f_2(x)(x - \bar{x})^2 \tag{2.3.4}$$

with $f_2(x)$ having a definite sign. Thus, it follows that

Case C: $f_2(x) > 0$

$$\begin{cases} \dfrac{dx}{dt} > 0, \text{if } x > \bar{x} \text{ or } x < \bar{x}, \end{cases} \tag{2.3.5}$$

Case D: $f_2(x) < 0$

$$\dfrac{dx}{dt} < 0, \text{if } x > \bar{x} \text{ or } x < \bar{x}, \tag{2.3.6}$$

Inspection of the flows along the x-axis, as indicated in Equations (2.3.2), (2.3.3), (2.3.5) and (2.3.6), shows that three types of behaviors are present: stable node, unstable node and saddle node, i.e.,

$$\longrightarrow\!\!\!\longrightarrow\bullet\longleftarrow\!\!\!\longleftarrow \quad \text{stable node } (S), \quad (2.3.7)$$

$$\longleftarrow\!\!\!\longleftarrow\bullet\longrightarrow\!\!\!\longrightarrow \quad \text{unstable node } (U), \quad (2.3.8)$$

$$\left.\begin{array}{c}\longrightarrow\!\!\!\longrightarrow\bullet\longrightarrow\!\!\!\longrightarrow\\ \longleftarrow\!\!\!\longleftarrow\bullet\longleftarrow\!\!\!\longleftarrow\end{array}\right\} \quad \text{saddle node } (SS), \quad (2.3.9)$$

where the saddle node is also called a 'semi-stable (SS)' fixed-point.

In more detail, the motion along the one-dimensional, x phase-space, for a single fixed-point, has one of the following three possibilities:

(a) If all trajectories approach the fixed-point, then $t \to \infty$, then it is a **stable node**.

(b) If all trajectories move away from the fixed-point, as $t \to \infty$, then it is an **unstable node**.

(c) If trajectories on one side of the fixed-point approach it as $t \to \infty$, while trajectories on the other side move away as $t \to \infty$, then it is a **saddle point**.

In addition, this general behavior continues to hold for a single, higher power zero of $f(x)$, i.e.,

$$\frac{dx}{dt} = f_1(x)(x - \bar{x})^n, \quad n > 1 \quad (2.3.10)$$

where $f_1(x)$ has a definite sign, i.e., $n =$ odd, produces results similar to Equations (2.3.2) and (2.3.3), while $n =$ even yields the conclusions stated in Equations (2.3.5) and (2.3.6).

2.4 TWO FIXED-POINTS

We now examine the case of a 1-dim system with three distinct fixed-points. Note that each fixed-point might correspond to a higher-order zero of $f(x)$. Below we draw the eight possible flows in the 1-dim phase-space. However, we indicate only half of the possible flows since the other four can be gotten by reversing the direction of the arrows and making the changes

$$S \to U, \quad U \to S, \quad SS \to SS. \tag{2.4.1}$$

$$A: \quad \underset{}{\longleftarrow \longleftarrow \bullet \overset{SS}{\longrightarrow \longrightarrow} \bullet \overset{SS}{\longrightarrow \longrightarrow}} \; x \tag{2.4.2}$$

$$B: \quad \underset{}{\longrightarrow \longrightarrow \bullet \overset{SS}{\longrightarrow \longrightarrow} \bullet \overset{S}{\longleftarrow \longleftarrow}} \; x \tag{2.4.3}$$

$$C: \quad \underset{}{\longrightarrow \longrightarrow \bullet \overset{S}{\longleftarrow \longleftarrow} \bullet \overset{U}{\longrightarrow \longrightarrow}} \; x \tag{2.4.4}$$

$$D: \quad \underset{}{\longrightarrow \longrightarrow \bullet \overset{S}{\longleftarrow \longleftarrow} \bullet \overset{SS}{\longleftarrow \longleftarrow}} \; x. \tag{2.4.5}$$

Note that if there are n distinct fixed-points, then there are 2^{n+1} flow diagrams.

Figure 2.1 gives plots of the solution behaviors, i.e., $x(t)$ vs t for the four cases (A, B, C, D) as indicated above.

2.5 LINEAR STABILITY

If the function $f(x)$ has simple zeros, then we can analyze the local stability properties in the neighborhood of each fixed-point. An advantage of carrying out such calculations is that it allows the determination of the various time scales of the system.

To proceed, let a system be modeled by the differential equation

$$\frac{dx}{dt} = f(x), \; x(0) = x_0 \text{ given}. \tag{2.5.1}$$

Let $\bar{x}$ be a simple zero of $f(x)$, i.e.,

$$f(\bar{x}) = 0. \tag{2.5.2}$$

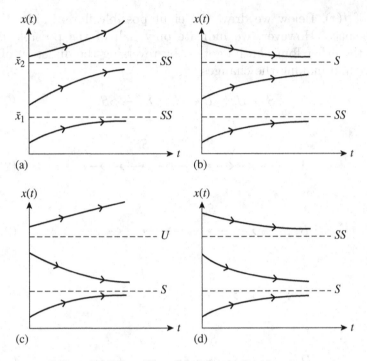

FIGURE 2.1 Solution plots, $x(t)$ vs t, for two fixed-points; only one haft is shown and they correspond to the 1-dim phase-space flows indicated in Equations (2.4.2)–(2.3.15).

For initial conditions in a small neighborhood of the fixed-point, $x(t) = \bar{x}$, we can write

$$x(t) = \bar{x} + \epsilon(t), \qquad (2.5.3)$$

where

$$|\epsilon(0)| \ll \bar{x}. \qquad (2.5.4)$$

Substituting Equation (2.5.3) into Equation (2.5.1) gives, upon doing a Taylor expansion, the result

$$\frac{dx}{dt} = f(\bar{x} + \epsilon) = f(\bar{x}) + \frac{df}{dx}\bigg|_{x=\bar{x}} \epsilon + O(\epsilon^2). \qquad (2.5.5)$$

If we retain only the linear term, then

$$\frac{d\epsilon}{dt} = R\epsilon, \quad \epsilon(0) = \epsilon_0, \qquad (2.5.6)$$

where R is the constant

$$R = \frac{df(x)}{dx}\bigg|_{x=\bar{x}}. \tag{2.5.7}$$

The solution to Equation (2.5.6) is

$$\epsilon(t) = \epsilon_0 e^{Rt}. \tag{2.5.8}$$

Observe that if $R > 0$, then $\epsilon(t)$ increases in magnitude, while for $R < 0, \epsilon(t)$ decreases to zero. From these results, we can determine the stability of the fixed-point, $x(t) = \bar{x}$,

$$R > 0 \Rightarrow x(t) = \bar{x}, \quad \text{locally unstable;} \tag{2.5.9}$$

$$R < 0 \Rightarrow x(t) = \bar{x}, \quad \text{locally stable.} \tag{2.5.10}$$

2.6 APPLICATIONS

2.6.1 Radioactive Decay

The differential equation modeling phenomena like simple radioactive decay are

$$\frac{dy}{dt} = -\lambda y, \quad y(0) = y_0, \text{given} \lambda > 0. \tag{2.6.1}$$

Since $y(t)$ is related to the amount of material undecayed at time t, this implies that all physical relevant solutions have the properties

$$y(t) \geq 0, \quad t > 0; \quad y(0) = y_0 \geq 0. \tag{2.6.2}$$

For this system $f(y) = -\lambda y < 0$, and the only fixed-point is $y(t) = \bar{y} = 0$. Note that

$$\frac{dy(t)}{dt} < 0, \quad y(t) > 0, \tag{2.6.3}$$

and it follows that the solution $y(t)$ begins at y_0 at $t = 0$ and monotonically decreases to zero. So, even if we could not solve Equation (2.6.1), we can conclude that a plot of $y(t)$ vs t ties the form presented in Figure 2.2.

28 ■ Introduction to Qualitative Methods for Differential Equations

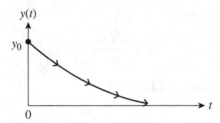

FIGURE 2.2   Qualitative behavior of the physical solutions to Equation (2.6.1).

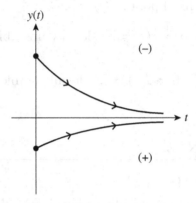

FIGURE 2.3   The right-hand solution plane for Equation (2.6.1) with two typical solution curves indicated.

Since $f(y) = -\lambda y$, we see that the fixed-point is a simple zero and the linear stability analysis yields $R = -\lambda$. Consequently, this procedure leads to the conclusion that the fixed-point $y(t) = \bar{y} = 0$ is linearly stable.

If we draw the full (right side) of the solution plane, then this result is illustrated in Figure 2.3. The ($\pm$) symbols indicate the 'sign' of dy/dt. Note that all solutions decrease in magnitude to zero, i.e.,

$$\lim_{t \to \infty} y(t) = 0. \tag{2.6.4}$$

However, Equation (2.6.1) can be solved exactly. To do so, let us rescale it by using the variables transformation

$$y(t) \to x(t) = \frac{y(t)}{y_0}, \quad t \to \bar{t} = \lambda t \tag{2.6.5}$$

to obtain

$$\frac{dx}{dt} = -x, \quad x(0) = 1, \tag{2.6.6}$$

having the solution

$$x(t) = e^{-\bar{t}} \text{ or } y(t) = y_0 e^{-\lambda t}. \tag{2.6.7}$$

Note that in going from $y(t)$ to the rescaled variable $x(t)$, the resulting differential equation and its solution do not explicitly depend on the parameter λ.

2.6.2 Logistic Equation

This differential equation is

$$\frac{dy}{dt} = ay - by^2, y(0) = y_0 > 0; a > 0, b > 0. \tag{2.6.8}$$

This equation may be rescaled by using the variable changes

$$y(t) \to x(t) = \frac{y(t)}{y_0}, t \to \bar{t} = at, \tag{2.6.9}$$

to obtain

$$\frac{dx}{dt} = x(1-x), \quad f(x) = x(1-x), \tag{2.6.10}$$

where the barred-t has been replaced by t in the last equation.

The derivative function $f(x)$ has two simple zeros, i.e.,

$$f(\bar{x}) = 0 \Rightarrow \bar{x}_1 = 0, \bar{x}_2 = 1. \tag{2.6.11}$$

Also, in physical application

$$x(0) = x_0 \geq 0, \quad x(t) \geq 0. \tag{2.6.12}$$

Figure 2.4 gives the solution plane for $x(t)$, indicates its two fixed-points, shows where the slope of $x(t)$ is positive or negative ($\pm$), and labels the stability of the fixed-points by S or U, i.e., stable or unstable.

Inspection of Figure 2.4 shows that for $0 < x_0 < 1$, $x(t)$ monotonically increases to the fixed-point value $\bar{x}_2 = 1$, and for $x_0 > 1, x(t)$ monotonically decreases to the fixed-point value $\bar{x}_2 = 1$. Thus, $\overline{X}_2$ is

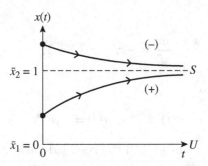

FIGURE 2.4 Major features of the solution plane for the logistic equation; see Equation (2.6.10).

a stable fixed-point. To summarize, all physical solutions have the property

$$\operatorname*{Lim}_{t\to\infty} x(t) = x(\infty) = \bar{x}_2 = 1. \qquad (2.6.13)$$

Equations (2.6.8) and (2.6.10) can be solved to find their exact solutions in terms of elementary functions; they are, respectively,

$$y(t) = \frac{y_0 K}{y_0 + (K - y_0)\, e^{-at}}, \quad K = \frac{a}{b}, \qquad (2.6.14)$$

and

$$x(t) = \frac{x_0}{x_0 + (1 - x_0)\, e^{-t}}. \qquad (2.6.15)$$

2.6.3 Gompertz Model

Another model with two fixed-points is that constructed by B. Gompertz. Its differential equation is

$$\frac{dy}{dt} = -ry\, \operatorname{Ln}\left(\frac{y}{K}\right), \quad y(0) = y_0 > 0, \qquad (2.6.16)$$

and

$$r > 0, \quad K > 0. \qquad (2.6.17)$$

For this case

$$\frac{dy}{dt} = f(y) \to f(y) = -ry\, \operatorname{Ln}\left(\frac{y}{k}\right), \qquad (2.6.18)$$

with
$$f(0) = 0, \quad f(k) = 0, \tag{2.6.19}$$
which gives the fixed-points
$$\bar{y}_1 = 0, \bar{y}_2 = K. \tag{2.6.20}$$
The following rescaling
$$y(t) \to x(t) = \frac{y(t)}{K}, t \to \bar{t} = rt, \tag{2.6.21}$$
gives the dimensionless equation
$$\frac{dx}{dt} = -x\text{Ln}(x), \quad x(0) = x_0 > 0. \tag{2.6.22}$$
where the bar over the t has been dropped. Equation (2.6.22) can be solved to give for $y(t)$ the answer
$$y(t) = K\left(\frac{y_0}{K}\right). \tag{2.6.23}$$
This result can be checked by evaluating $y(0)$ and $y(\infty)$, i.e.,
$$y(0) = K\left(\frac{y_0}{K}\right) = y_0, \tag{2.6.24}$$
$$y(\infty) = \lim_{t \to \infty} y(t) = K = \bar{y}_2. \tag{2.6.25}$$

It should be indicated that the Gompertz equation has a solution plane representation that is exactly the same as the logistic equation; see Figure 2.4. This is a consequence of both differential equations having just two fixed-points with the same stability properties.

The Gompertz differential equation, for most students and professionals, would be difficult to integrate. However, it has been shown that the qualitative approach allows an easy and direct determination of the general features of solutions to this equation.

2.6.4 Draining a Tank

There exists several physical systems that can be modeled by a first-order different equation having the form
$$\frac{dy}{dt} = -k\sqrt{y}, y(t) > 0, y(0) = y_0 > 0. \tag{2.6.26}$$

32 ■ Introduction to Qualitative Methods for Differential Equations

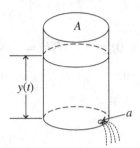

FIGURE 2.5 Fluid following out of a tank from the small hole at the bottom with area a and constant cross-section A. The height of the fluid at time $t > 0$ is $y(t)$.

In particular, it can represent the draining of a tank filled with water or some other freely flowing liquid.

(By 'freely flowing liquid', we mean such fluids having sufficiently small viscosities as water.)

The geometry of the system is shown in Figure 2.5. It consists of a cylindrical tank of constant cross-sectional area A, and at the bottom a small circular hole of area a. We assume

$$a<<A. \tag{2.6.27}$$

If the tank is initially filled to a height of y_0, we are interested in how the height of the liquid surface changes with time, i.e., the solution of the mathematically problem given in Equation (2.6.26).

It should be noted that application of the fundamental laws of physics to this system, along with certain physically valid approximations, gives the following value for the parameter k in Equation (2.6.26)

$$k = \sqrt{2g}\left(\frac{a}{A}\right), \tag{2.6.28}$$

where g is the acceleration due to gravity at the surface of the earth.

While Equation (2.6.26) is solvable, the general changes in the height, $y(t)$, as a function of time can be easily determined by considering an experiment corresponding to a cylindrical tank with a small hole at the bottom. If the tank is filled to a height y_0 and the hole at the bottom is unplugged, then the surface height will decrease with time and at some definite finite future time the tank will be empty. This analysis implies that $y(t)$, the solution to Equation (2.6.26), should look like the behavior shown in Figure 2.6.

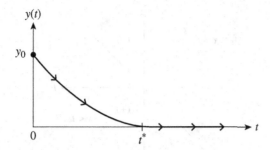

FIGURE 2.6 Plot of $y(t)$ vs t. The value of $y(t)$ becomes zero at a finite time $t = t^*$.

It should also be clear that at $t = t^*$, the function $y(t)$ is continuous with a continuous derivative since

$$y(t^*) = 0, \quad \frac{dy(t^*)}{dt} = 0, \quad \frac{d^2y(t^*)}{dt^2} = k^2 > 0. \tag{2.6.29}$$

Again, note that Equation (2.6.26) can be easily solved since it is a separable, first-order, ordinary differential equation. Its solution is the piece-wise continuous function

$$y(t) = \begin{cases} \left(\sqrt{y_o} - \frac{kt}{2}\right)^2, & \text{for } 0 < t \leq t^*; \\ 0, & \text{for } t \geq t^*, \end{cases} \tag{2.6.30}$$

where

$$t^* = \frac{2\sqrt{y_0}}{K}. \tag{2.6.31}$$

There are two further comments. First, it should be clear that the solutions we seek hold only for $y \geq 0$. Negative solutions are physically meaningless. Second, $y(t) = 0$ is a physically meaningful solution in that it corresponds to the tank being empty. This also follows from an inspection of Equation (2.6.26), which tells us that $y(t) = \bar{y} = 0$ is a fixed-point.

2.6.5 $f(x, t)$ Depends on t

Up to now, we have only investigated first-order autonomous differential equations, i.e., those equations for which the function 'f' depended on x, but not t. However, let's see what can be done when

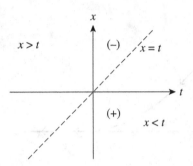

FIGURE 2.7 Regions in the $x-t$ plane where the slopes of the solutions to Equation (2.6.32) have definite signs, except along the line $x = t$.

'f does depend on t. Perhaps the simplest example of this is the linear equation

$$\frac{dx}{dt} = t - x, \quad x(0) = x_0. \tag{2.6.32}$$

First, observe that in the $x-t$ plane, the slope of $x(t)$ is zero on the line $x^{(0)}(t) = t$. Note that $x^{(0)}(t) = t$ is not a solution to Equation (2.6.32).

Second, we also have the slope negative at all points in the $x-t$ plane where $x(t) > t$ and positive at points for which $x(t) < t$. This is indicated in Figure 2.7.

Third, using this information, we can sketch the corresponding solution curves in the $x-t$ plane. These results are presented in Figure 2.8 for several different solution trajectories.

For this particular differential equation, we can solve it to obtain its exact solution, which is

$$x(t) = Ae^{-t} - 1 + t, \tag{2.6.33}$$

where A is an arbitrary constant whose value is determined by the initial condition on $X(t)$. For example, if

$$X(0) = x_0 \text{ is given}, \tag{2.6.34}$$

then

$$A = x_0 + 1 \tag{2.6.35}$$

and

$$x(t) = (x_0 + 1)e^{-t} - 1 + t. \tag{2.6.36}$$

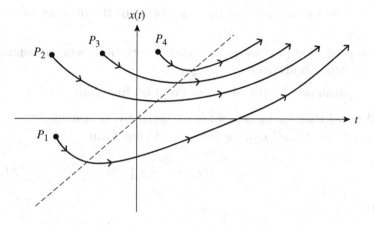

FIGURE 2.8 Sketchs of solutions to the differential Equation (2.6.32).

Comparison of this analytic solution to the qualitative features shown by the $x(t)$ vs t curves in Figure 2.8 shows that they are consistent with each other. However, the details of the general analytic behavior of such solutions, as presented in Figure 2.8, cannot be obtained from our qualitative examination. For example, from Figure 2.8, it may be concluded that all solutions increase in magnitude when $|t|$ increases, for sufficiently large $|t|$. However, the qualitative procedure cannot provide the details as to what this increase is. But the analytical solution does contain these answers, i.e., from Equation (2.6.36)

$$x(t) = \begin{cases} (x_0 + 1)\,e^{-t}, & \text{for } t < 0, \text{large}; \\ (-1 + t), & \text{for } t > 0, \text{large}. \end{cases} \qquad (2.6.37)$$

2.6.6 Spruce Budworm Population Model

One of the most important mathematical models in single population dynamics is the so-called 'spruce budworm model (SBM)', which takes the form

$$\frac{dx}{dt} = r_0 x\left(1 - \frac{x}{K}\right) - \frac{\rho x}{x + A}, \qquad (2.6.38)$$

where

r_0 = growth rate of spruce budworm for small populations,

K = carrying capacity of the budworms in the absence of predators,

ρ = predation parameter, associated with birds who consume the spruce budworms,

A = parameter in the bird consumption function.

All three of these parameters are assumed to be positive.

Equation (2.6.38) can be rewritten to the form

$$\frac{dx}{dt} = x[r(x) - p(x)], \qquad (2.6.39)$$

where

$$r(x) = r_0\left(1 - \frac{x}{K}\right), \; p(x) = \frac{\rho}{x + A}. \qquad (2.6.40)$$

Inspection of Equation (2.6.39) shows that there is always a fixed-point at

$$\bar{x}_1 = 0. \qquad (2.6.41)$$

Consequently, any other fixed-points must come from the term

$$r(x) - p(x) = 0 \qquad (2.6.42)$$

or

$$x^2 + (A - K)x + \left(\frac{K\rho}{r_0} - A\right) = 0. \qquad (2.6.43)$$

While the last equation is just a quadratic algebraic equation, it depends on four parameters, (r_0, A, K, ρ) and would be very difficult to analyze. However, a way to get around this issue is to understand that an equilibrium or fixed-point will occur whenever the two curves, $r(x)$ and $p(x)$, intersect. Figure 2.9 shows that three possibilities exist:

(i) $r(x)$ and $p(x)$ do not intersect. This means that there is only the fixed-point at $\bar{x}_1 = 0$.

(ii) $r(x)$ and $P(x)$ intersect at two points. This means that in addition to the fixed-point, $\bar{x}_1 = 0$, there are two real and positive fixed-points at $\bar{x}_2 > 0$ and $\bar{x}_3 > 0$.

(iii) $r(x)$ and $p(x)$ intersect, but one intersection occurs at a negative value of x, with the other occurring at $\bar{x}_1 = 0$ and $\bar{x}_3 > 0$.

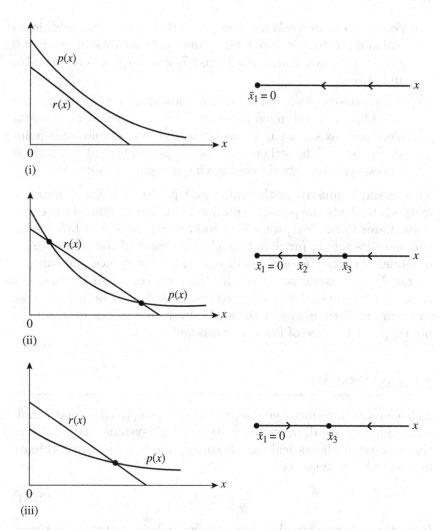

FIGURE 2.9 Plots of $p(x)$ and $r(x)$ vs X for the spruce budworm model. The items on the right side are the corresponding flow diagrams.

In ecological terms, the explanation goes as follows:

(a) When predation is very large, $p(x) > r(x)$, then the two curves never intersect and $dx/dt < 0$ for all $x > 0$. Consequently, the spruce budworm population goes to extinction and settles in the $\bar{x}_1 = 0$ state of stable equilibrium. Note that $\bar{x}_1$ is a stable fixed-point.

(b) For a somewhat weakened predation, there can be two additional real and positive fixed-points: $\bar{x}_2$ and $\bar{x}_3$. This case still has $\bar{x}_1 = 0$ as a stable fixed-point, but $\bar{x}_2$ and $\bar{x}_3$ are, respectively, unstable and stable.

(c) If the consumption of the spruce budworms is very weak, then in addition to the fixed-point at $\bar{x}_1 = 0$, which is now unstable, there exists a single, real, positive and stable fixed-point at $\bar{x}_3 > 0$. The other fixed-point is unphysical because it corresponds to a flxed-point having a negative value.

This example illustrates the value and power of using qualitative methods to study the possible solution behaviors of differential equations. Later in the text, we will introduce the concept of bifurcation and show its role in providing insight into some of the issues arising in models such as the one just discussed for the spruce budworm.

Finally, an examination of the three outcomes shows that the rightmost fixed-point is always stable. This has to be a prior case since any realistic model of spruce budworm dynamics must always predict the existence of finite worm populations.

2.7 DISCUSSION

This chapter introduced geometric techniques applied to mathematical models of one-dimensional systems. Such systems are defined as those whose mathematical models correspond to first-order, ordinary differential equations, i.e.,

$$\frac{dx}{dt} = f(x). \tag{2.7.1}$$

In Section 2.5, we have also given reasons to believe that (sometimes) these techniques can be extended to nonautonomous systems whose mathematical model is represented by equations of the type

$$\frac{dx}{dt} = g(x, t). \tag{2.7.2}$$

This can always occur if $g(x, t)$ has one of the following structures

$$g(x, t) = f_1(x) - f_2(t), \quad g(x, t) = f_1(x) f_2(t). \tag{2.7.3}$$

Our general conclusion is that even in the absence of exact solutions to Equations (2.7.2) and (2.7.3), important information on the properties of their solutions may be determined from the applicant of qualitative methods related to the geometric features of the 1-dim flow diagrams and the associated $x-t$ solution planes.

PROBLEMS

Analyze the following differential equations:

(1) $\frac{dx}{dt} = -\lambda_1 x - \lambda_2 x^{\frac{1}{3}}; \quad \lambda_1 > 0, \lambda_2 > 0$

(2) $\frac{dx}{dt} = t - x^2$

(3) $\frac{dx}{dt} = rx^2\left(1 - \frac{x}{k}\right); \quad r > 0, k > 0$

(4) $\frac{dx}{dt} + x^3 = \sin t$

In particular

- Determine the value of any fixed-points;
- Determine the stability of any found fixed-points;
- Sketch the x flow-fields and/or the trajectors in the $x-t$ solution plane.

Try to think of possible physical systems corresponding to each of these differential equations. Can any of these equations be solved analytically in terms of known functions? If so, then solve for these solutions.

NOTES AND REFERENCES

Sections 2.2, 2.3 and 2.4: The materials in these sections are based on my book

(1) R. E. Mickens, *Mathematical Methods for the Natural and Engineering Sciences*, 2nd Edition (World Scientific, Singapore, 2017). See Sections 4.1, 4.2 and 4.3.

Section 2.5: References for the models introduced in this section are

(2) Radioactive decay: W. Loveland, D. Morrissey, and G. T. Seaborg, *Modern Nuclear Chemistry* (Wiley-Interscience, New York, 2006).

(3) Logistic equation: J. D. Murray, *Mathematical Biology* (Springer-Verlag, Berlin, 1989).

(4) Gompertz model: C. P. Winor, The Gompertz curve as a growth curve, *Proceedings of the National Academy of Science*, Vol. 18 (1932), 1–8.

(5) Draining a tank: T.-R. Hsu, *Applied Engineering Analysis* (Wiley, New York, 2018). See Sections 7.4.1 and 7.4.2.

(6) Spruce budworm model: D. Ludwig, D. D. Jones, and C. S. Holling, Qualitative analysis of insect outbreak systems: The spruce budworm and forest, *Journal of Animal Ecology*, Vol. 47 (1978), 315–332.

See also Ref. [3] above.

CHAPTER 3

Two-Dimensional Dynamical Systems

3.1 INTRODUCTION

Many of the physical systems of interest can be modeled by a single, second-order, differential equation having the form

$$\frac{d^2x}{dt^2} = F\left(x, \frac{dx}{dt}\right). \qquad (3.1.1)$$

This expression can be rewritten as two coupled first-order equations

$$\frac{dx}{dt} = y, \frac{dy}{dt} = F(x, y). \qquad (3.1.2)$$

If we take the initial conditions of Equation (3.1.1) to be $x(0) = x_0$ and $dx(0)/dt = x'_0$, then the corresponding initial conditions for Equation (3.1.2) are $x(0) = x_0$ and $y_0 = y(0) = x'_0$.

The main purpose of this chapter is to examine a somewhat more general pair of differential equations than that expressed in Equation (3.1.2), i.e.,

$$\frac{dx}{dt} = f(x, y), \frac{dy}{dt} = g(x, y). \qquad (3.1.3)$$

Since $f(x, y)$ and $g(x, y)$ do not depend explicitly on the independent variable, t, these differential equations are called autonomous. However, non-autonomous equations may always be transformed into autonomous ones by simple renaming of the independent variable.

For example, consider

$$\frac{dx}{dt} = F_1(x, y, t), \frac{dy}{dt} = G_1(x, y, t), \qquad (3.1.4)$$

DOI: 10.1201/9781003422419-4

then an equivalent set of equations is

$$\frac{dx}{dt} = F_1(x, y, z), \frac{dy}{dt} = G_1(x, y, z), \frac{dz}{dt} = 1. \quad (3.1.5)$$

In the next section, we introduce a number of definitions of concepts central to the study of 2-dim dynamical systems. This is followed by Section 3.3 in which we discuss some very general features of the trajectories in the 2-dim phase-plane. Section 3.4 demonstrates how phase diagrams for the trajectories are to be done. This is followed by Sections 3.5 and 3.6 where, respectively, linear stability analysis is presented, followed by a brief discussion of the local behavior of the trajectories near fixed-points for nonlinear systems. Section 3.7 gives working illustrations of how one can apply 2-dim phase-space techniques to determine and analyze the qualitative behavior of a large number of dynamical systems.

3.2 DEFINITIONS

3.2.1 2-Dim Dynamical System

A 2-dim dynamical system is one whose mathematical model is

$$\begin{cases} \frac{dx}{dt} = P(x, y), \frac{dy}{dt} = Q(x, y), \\ x(0) = x_0, y(0) = y_0 \end{cases} \quad (3.2.1)$$

It is assumed that $P(x, y)$ and $Q(x, y)$ have properties such that unique solutions exist for the problems of interest.

The phase-plane is the space (x, y), and the trajectories are curves, $y = y(x)$, obtained by solving the first-order differential equation

$$\frac{dy}{dx} = \frac{Q(x, y)}{P(x, y)}. \quad (3.2.2)$$

This equation follows from noting that since

$$x = x(t), \quad y = y(t), \quad (3.2.3)$$

then

$$\frac{dy(x)}{dt} = \frac{dy}{dx}\frac{dx}{dt} \quad (3.2.4)$$

and

$$\frac{dy}{dx} = \frac{dy/dt}{dx/dt} = \frac{Q(x, y)}{P(x, y)}. \quad (3.2.5)$$

3.2.2 Fixed-Points

The fixed-points are constant or equilibrium solutions to Equation (3.2.1)

$$x(t) = \bar{x}, y(t) = \bar{y}, \qquad (3.2.6)$$

i.e., they correspond to time-independent solutions, and therefore are solutions to the simultaneous equations

$$P(\bar{x}, \bar{y}) = 0, \quad Q(\bar{x}, \bar{y}) = 0. \qquad (3.2.7)$$

It should be observed that in most cases only the real solutions have 'physical interpretations'.

3.2.3 Nullclines

There are two important curves in the (x, y) plane that while not, in general, being solution trajectories, play critical roles in constructing the behavior of solution trajectories. With respect to Equation (3.2.2), these curves are the x-nullcline and the y-nullcline, and they are defined as follows:

(i) The x-nullcline is the curve along which $dy/dx = \infty$.

(ii) The y-nullcline is the curve along which $dy/dx = 0$.

Note that each nullcline may consist of several disjointed segments.

If these curves are denoted, respectively, by $y_\infty(x)$ and $y_0(x)$, then their functional forms can be calculated by solving the equations

$$P[x, y_\infty(x)] = 0, Q[x, y_0(x)] = 0. \qquad (3.2.8)$$

With regard to the nullclines, it is important to be aware of the following points:

(a) $y = y_\infty(x)$ and $y = y_0(x)$ are not in general solution to Equation (3.2.2).

(b) The places where these two curves intersect are the fixed-points of the system.

(c) The places where either nullcline intersects with itself have no particular significance.

(d) The nullclines divide the phase-plane into many open domains. The boundaries of these domains are the nullclines. Moreover,

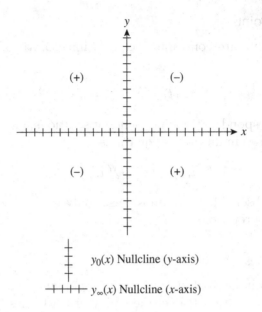

FIGURE 3.1 Nullclines for $dy/dx = -(x/y)$.

in each separate domain, the 'sign' of the derivative, dy/dx, is constant, i.e., dy/dx is bounded and is either positive or negative. The only way the derivative can change its sign is to cross from one open domain to another.

A simple illustration of these concepts is shown in Figure 3.1 for the system

$$\frac{dx}{dt} = y, \quad \frac{dy}{dt} = -x, \tag{3.2.9}$$

for which

$$\frac{dy}{dx} = -\left(\frac{x}{y}\right); \quad \text{fixed-point } (\bar{x}, \bar{y}) = (0, 0) \tag{3.2.10}$$

and nullclines

$$y_\infty(x) : y = 0, \text{ the } x\text{-axis}; \tag{3.2.11}$$

$$y_0(x) : x = 0, \text{ the } y\text{-axis}. \tag{3.2.12}$$

Note that for this simple example, the coordinate axes are the nullclines, and they divide the x–y plane into four domains, in which the derivative, dy/dx, has the indicated constant sign. Further, the two nullclines intersect at the origin, which as expected is the location of the only fixed-point $(\bar{x}, \bar{y}) = (0,0)$.

3.2.4 First-Integral and Symmetry Transformations

A first-integral is defined to be a general solution to Equation (3.2.2). If we denote it by $I(x, y)$, then

$$I(x, y) = C, \qquad (3.2.13)$$

where C is an arbitrary integration constant, determined by the initial conditions, i.e.,

$$I(x_0, y_0) = C. \qquad (3.2.14)$$

In general, this means that $y = y(x)$ can only be obtained implicitly by means of the first-integral. For a specific value of C, say C_1, the curve

$$y = y(x, c_1) \qquad (3.2.15)$$

in the x–y plane is called a level curve. Observe that different values of C give rise to different level curves. The totality of level curves is the level set of the system.

In many cases, $I(x, y)$ will be invariant under a change of coordinates. If this occurs, then the system has a symmetry. Examples of possible elementary symmetries include

(i) $x \to -x, y \to y$: reflection in the y-axis;

(ii) $x \to x, y \to -y$: reflection in the x-axis;

(iii) $x \to -x, y \to -y$: inversion through the origin.

It will be seen later in this book that the existence of first-integrals and/or symmetry transformations plays very important roles in the construction and interpretation of the paths of trajectories in the 2-dim $x - y$ plane.

3.3 GENERAL FEATURES OF TRAJECTORIES

Trajectories in the $x - y$ plane for a dynamical system, generally, will have one of the following behaviors:

(1) A trajectory may approach a fixed-point as $t \to +\infty$, i.e.,

$$\lim_{t \to +\infty} \begin{pmatrix} x(t) \\ y(t) \end{pmatrix} = \begin{pmatrix} \bar{x} \\ \bar{y} \end{pmatrix}. \qquad (3.3.1)$$

(2) A trajectory may become unbounded as $t \to \infty$, i.e.,

$$\lim_{t \to +\infty} \begin{pmatrix} x(t) \\ y(t) \end{pmatrix} = \begin{pmatrix} \infty \\ \infty \end{pmatrix}.$$

(3) If a trajectory begins at a fixed-point, it remains there for all $t > 0$.

(4) A trajectory may be a simple (i.e., nonintersecting) closed curve.

(5) A trajectory may approach a closed curve as $t \to +\infty$, or a trajectory that begins in the neighborhood of a closed curve may move away from it as $t \to +\infty$.

See Figure 3.2 for graphic representations of these cases.

If we wish to include more detail in our understanding of the behavior of the trajectories, a better understanding of the local behavior of the trajectories in the neighborhood of fixed-points is needed. The following possibilities may occur:

(1) Stable node: All trajectories approach the fixed-point along non-spiraling curves, i.e., as $t \to +\infty$, the trajectories are asymptotic to straight lines.

(2) Unstable node: All trajectories leave a neighborhood of the fixed-point along non-spiraling curves.

(3) Stable spiral node: All trajectories approach the fixed-point along spiral curves as $t \to +\infty$.

(4) Unstable spiral node: All trajectories leave a neighborhood of the fixed-point along spiral curves as $t \to +\infty$.

(5) Saddle point: Trajectories initially move toward the fixed-point and then move away. There are two trajectories that move toward the fixed-point and end on it. Likewise, there are two trajectories that originate at the fixed-point, but move away from it. These respective trajectories are called the stable and unstable manifolds of the saddle point.

(6) Center: All neighboring trajectories form closed curves about a fixed-point.

Sketches of these six behaviors are given in Figure 3.3.

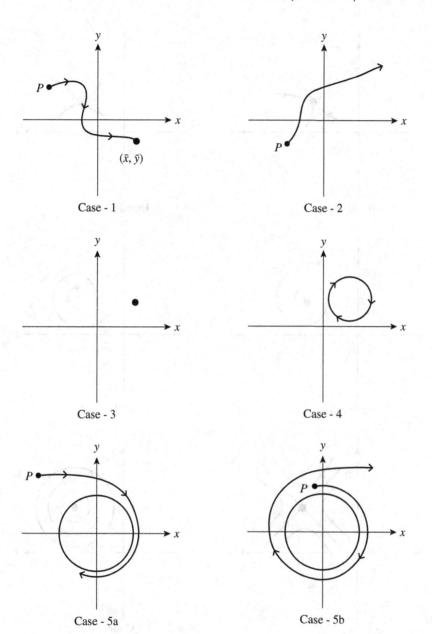

FIGURE 3.2 Trajectory possibilities.

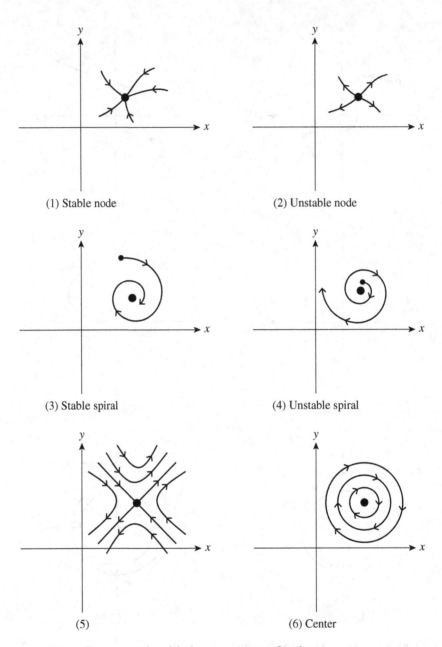

FIGURE 3.3 Trajectory local behaviors near a fixed-point.

3.4 CONSTRUCTING PHASE-PLANE DIAGRAMS

Given a 2-dim dynamic system, the behavior of its trajectories in the x-y plane can generally be determined by the following steps.

(A) Calculate the location of the real fixed-points of the system

$$\frac{dx}{dt} = f(x, y), \quad \frac{dy}{dt} = g(x, y), \qquad (3.4.1)$$

by solving the pair of equations

$$f(\bar{x}, \bar{y}) = 0, \, g(\bar{x}, \bar{y}) = 0. \qquad (3.4.2)$$

Denote these fixed-points by

$$\left\{ \bar{x}^{(i)}, \bar{y}^{(i)} \; : \; i = 1, 2, ..., I \right\}, \qquad (3.4.3)$$

I is the total number of real fixed-points.

(B) Calculate the x-nullcline by solving for $y_\infty(x)$ the equation

$$f[x, y_\infty(x)] = 0. \qquad (3.4.4)$$

(C) Calculate the y-nullcline by solving for $y_0(x)$ the equation

$$g[x, y_0(x)] = 0. \qquad (3.4.5)$$

(D) Draw on the $x - y$ plane the (real) fixed-points and x- and y-nullclines. Note that there should not be any fixed-points that do not lie at the intersections of the x- and y-nullclines.

(E) The x- and y-nullclines will divide the phase-plane into several open domains. In each separate domain, the sign of dy/dx is either all positive or all negative. Further, dy/dx is bounded in each domain. Determine the sign of the derivative for each of the domains.

(F) Pick an 'appropriate point' on the plane and sketch carefully the trajectory passing through this point. Repeat this for a number of other points until the general flow of the trajectories in the plane becomes clear.

If the system equations or the trajectory equation

$$\frac{dy}{dt} = \frac{g(x, y)}{f(x, y)} \qquad (3.4.6)$$

is invariant under certain coordinate transformations, then use this information to and in the sketching of the trajectories.

(G) Following the above steps will often provide a unique geometrically structure for the flow of the trajectories. When there are ambiguities, it may be helpful to examine in more detail the local behavior of the fixed-points in their neighborhoods.

The next section presents a brief discussion of the use (when applicable) of the concept of linear stability analysis.

3.5 LINEAR STABILITY ANALYSIS

Consider a 2-dim, linear, autonomous system

$$\begin{cases} \frac{dx}{dt} = ax + by, \\ \frac{dy}{dt} = cx + dy, \end{cases} \quad (3.5.1)$$

where (a, b, c, d) are real constants. In general, the fixed-point is located at $(\bar{x}, \bar{y}) = (0, 0)$. We can rewrite these two, coupled differential equations in the matrix form

$$\frac{d\overline{X}}{dt} = A\overline{X}, \quad (3.5.2)$$

where

$$\overline{X} = \begin{pmatrix} x \\ y \end{pmatrix}, \quad A = \begin{pmatrix} a & b \\ c & d \end{pmatrix}. \quad (3.5.3)$$

Let λ_1 and λ_2 be the eigenvalues of matric A, i.e., they are solutions to the equation

$$\det(A - \lambda I) = \lambda^2 - (a + d)\lambda - (ad - bc) = 0, \quad (3.5.4)$$

where I is the 2 × 2 unit matrix. If $\lambda_1 \neq \lambda_2$, then associated with each of these eigenvalues is an eigenvector, i.e.,

$$A V_i = \lambda_i V_i, \quad i = (1, 2). \quad (3.5.5)$$

Thus, the general solution can be expressed as

$$X(t) = c_1 V_1 e^{\lambda_1 t} + c_2 V_2 e^{\lambda_2 t}, \quad (3.5.6)$$

where (c_1, c_2) arbitrary constants. A_N examination of Equation (3.5.6) provides the following conclusions:

(a) If λ_1 and λ_2 are both **real and negative**, then all trajectories approach the fixed-point as $t \to +\infty$ and the fixed-point is a **stable node**.

(b) If λ_1 and λ_2 are both **real and positive**, then all trajectories move away from the fixed-point as $t \to +\infty$ and the fixed-point is an **unstable node**.

(c) If λ_1 and λ_2 are both real, but λ_1 is positive and λ_2 is negative, then the trajectories approach in the direction of V_2 and move away in the direction of V_1. For this case, the fixed-point is a **saddle point**.

(d) If the λ are complex conjugates, i.e., $\lambda_1 = \lambda_2^*$, then if then the fixed-point is a **spiral point**. For $\text{Re}\,\lambda_1 = \text{Re}\,\lambda_2 < 0$, the trajectories spiral toward the fixed-point, while for $\text{Re}\,\lambda_1 = \text{Re}\,\lambda_2 > 0$, the trajectories away from the fixed-point. These cases correspond, respectively, to **stable** and **unstable spiral points**.

$$\text{Re}\,\lambda_1 = \text{Re}\,\lambda_2 \neq 0, \qquad (3.5.7)$$

(e) If λ_1 and λ_2 are purely imaginary, i.e., then the vector $X(t)$ describes a closed curve and the motion or solution is periodic.

$$\text{Re}\,\lambda_1 = \text{Re}\,\lambda_2 = 0, \qquad (3.5.8)$$

Finally, with these results and classifications, the general structure of the trajectories in the x–y plane can be easily determined. However, since our systems are generally nonlinear, the above results may not hold. In the next section, we resolve some of these issues.

3.6 LOCAL BEHAVIOR OF NONLINEAR SYSTEMS

Consider the following nonlinear system

$$\begin{cases} \dfrac{dx}{dt} = ax + by + F_1(x,y), \\ \dfrac{dy}{dt} = cx + dy + G_1(x,y), \end{cases} \qquad (3.6.1)$$

where

$$\lim_{\substack{x \to 0 \\ y \to 0}} \left\{ \begin{array}{c} \frac{F_1(x,y)}{r(x,y)} \\ \frac{G_1(x,y)}{r(x,y)} \end{array} \right\} = \begin{pmatrix} 0 \\ 0 \end{pmatrix} \qquad (3.6.2)$$

and
$$r^2 = x^2 + y^2. \tag{3.6.3}$$

(Note that we are still assuming that the fixed-point is located at $(\bar{x}, \bar{y}) = (0,0)$.) Our task is to determine how the stability of the fixed-point changes as we go from the linear system, Equation (3.5.1), to the nonlinear system, Equation (3.6.1).

The fixed-point at $(\bar{x}, \bar{y}) = (0,0)$ is said to be stable if the initial point (x_0, y_0) is sufficiently close to the fixed-point such that $x(t)$ and $y(t)$ remain close to $(\bar{x}, \bar{y}) = (0,0)$ for all $t > 0$. This can be formulated for a general fixed-point by using the following vector representation:

$$\overline{X}(t) = \begin{pmatrix} x(t) \\ y(t) \end{pmatrix}, \overline{X}_0(0) = \begin{pmatrix} x_0 \\ y_0 \end{pmatrix}, \overline{X} = \begin{pmatrix} \bar{x} \\ \bar{y} \end{pmatrix} \tag{3.6.4}$$

and

$$|\overline{X}_0 - \overline{X}| = \sqrt{(x_0 - \bar{x})^2 + (y_0 - \bar{y})^2}. \tag{3.6.5}$$

The fixed-point $\bar{x}$ is stable provided that for each $\epsilon > 0$, there exists a δ such that

$$|\bar{x}_0 - \bar{x}| < \delta \Rightarrow |\bar{x}(t) - \bar{x}| < \epsilon, t > 0. \tag{3.6.6}$$

A fixed-point is unstable if it is not stable.

A fixed-point, $\overline{X}$, is **asymptotically stable** if it is stable and every trajectory that starts sufficiently close to it approaches it as $t \to \infty$, i.e.,

$$|\bar{x}(t) - \bar{x}| < \delta \Rightarrow \lim_{t \to \infty} x(t) = \bar{x}. \tag{3.6.7}$$

For the nonlinear system given in Equation (3.6.1), the following theorem provides information on the stability properties of the fixed-point $(\bar{x}, \bar{y}) = (0,0)$, provided $ad - bc \neq 0$.

Theorem:

Let λ_1 and λ_2 be the eigenvalues associated with the matrix of the linear coefficients of Equation (3.6.1), i.e.,

$$\det(A - \lambda I) = \det \begin{vmatrix} a - \lambda & b \\ c & d - \lambda \end{vmatrix}$$
$$= \lambda^2 - (a + d)\lambda + (ad - bc)$$
$$= (\lambda - \lambda_1)(\lambda - \lambda_2).$$

TABLE 3.1 Classification of the Fixed-Points for Nonlinear Systems

Eigenvalues	Type of Fixed-Point and Stability
$\lambda_1 > \lambda_2 > 0$	Unstable node
$\lambda_1 < \lambda_2 < 0$	Stable node
$\lambda_1 = \lambda_2 > 0$	Unstable node or spiral point
$\lambda_1 = \lambda_2 < 0$	Stable node or spiral point
$\lambda_1 < 0 < \lambda_2$	Saddle point (unstable)
$\lambda_1 = \lambda_2^* = a + bi\ (a > 0)$	Unstable spiral point
$\lambda_1 = \lambda_2^* = a + bi\ (a < 0)$	Stable spiral point
$\lambda_1 = \lambda_2^* = bi$	Stable center or stable or unstable spiral point

Then the stability properties of the fixed-point for the nonlinear system of Equation (3.6.1) are as given in Table 3.1.

Examination of Table 3.1 indicates that if a linear stability analysis of a fixed-point is a center, then the corresponding nonlinear system may not have it as a center. It may turn out to be a stable or unstable spiral point. Thus, centers are very fragile fixed-points.

In the next section, we use these results and related geometrical techniques to sketch the flow of a variety of 2-dim systems. While some of these equations have known exact analytical solutions, our purpose is to examine them as if we did not know the actual solutions and see how much information can be obtained on the qualitative properties of the solutions. This is clearly the situation that confronts most researchers or even students when given arbitrary differential equations.

3.7 EXAMPLES

3.7.1 Harmonic Oscillator

The harmonic oscillator is modeled by the equation

$$\frac{d^2x}{dt^2} + x = 0,\ x(0) = x_0,\ \frac{dx(0)}{dt} = v_0, \qquad (3.7.1)$$

where x_0 and v_0 are the respective location and velocity at $t = 0$. In a system form, this equation can be represented as

$$\frac{dx}{dt} = y,\ \frac{dy}{dt} = -x,\ \text{where}\, x(0) = x_0, y(0) = y_0. \qquad (3.7.2)$$

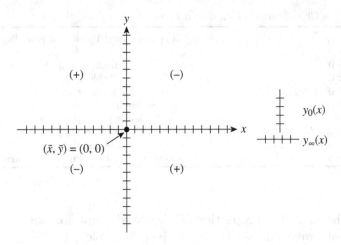

FIGURE 3.4 Basic features of the phase-plane for the harmonic oscillator.

Note that the system equations could also be written as

$$\frac{dx}{dt} = -y, \quad \frac{dy}{dt} = x, \text{ where } x(0) = x_0, y(0) = v_0. \tag{3.7.3}$$

Examination of Equation (3.7.2) indicates that there is only one fixed-point and it is

$$\bar{x} = 0, \bar{y} = 0. \tag{3.7.4}$$

The equation determining the trajectories in the $x-y$ phase-space is

$$\frac{dy}{dx} = -\frac{x}{y}, \tag{3.7.5}$$

and the associated nullclines are

$$\begin{cases} \frac{dy}{dx} = 0 : x = 0 \text{ or } y_0(x) \text{ is the } y\text{-axis;} \\ \frac{dy}{dx} = \infty : y = 0 \text{ or } y_\infty(x) \text{ is the } x\text{-axis.} \end{cases} \tag{3.7.6}$$

Observe that the nullclines intersect at the origin, which is where the only fixed-point is located, i.e., $(\bar{x}, \bar{y}) = 0$. Figure 3.4 shows this information and the signs of the derivative in the four domains created by the nullclines.

Equation (3.7.5) can be integrated to give

$$I(x,y) = x^2 + y^2 = x_0^2 + y_0^2 = \text{constant}. \tag{3.7.7}$$

Note that the first-integral, $I(x,y)$, is invariant under the following coordinate transformations:

$$T_1 \; : \; x \to -x, y \to y; \text{reflection in the } y\text{-axis.}$$

$$T_2 \; : \; x \to x, y \to -y; \text{ reflection in the } x\text{-axis.}$$

$$T_3 = T_1 T_2 = T_2 T_1 \; : \; \text{ inversion through the origins.}$$

$$x \to -x, \quad y \to -y.$$

We now use these transformations to construct a typical trajectory in the 2-dim $x-y$ plane for the harmonic oscillator. This is pictorially illustrated in Figure 3.5. In more detail, we carry out the following steps:

(1) In the upper left diagram, the fixed-point and four domains are indicated, along with the signs of the dy/dx. The x- and y-axes are, respectively, the $y_\infty(x)$ and $y_0(x)$ nullclines.

(2) Starting at point-1, on the y-axis, we obtain a curve from point-1 to point-2, which is on the x-axis. Since the derivative must be zero at point-1, negative in the first quadrant, and infinite on the x-axis, the curve joining points -1 and -2 must have the shape indicated in the upper right figure.

(3) If this curve is now reflected in the x-axis, i.e., applying T_2, we obtain the curve shown in the lower right figure.

(4) Applying T_1 to the curve labeled by $1-2-3$, we obtain the full trajectory as depicted in the lower left figure and this is a closed curve.

Comments

(i) Since T_1 and T_2 are symmetry transformations, it is clear that point-1 and point-3 are the same exact distances from the origin, with the same holding for point-2 and point-4. This also means that point-1 and point-5 coincide (as drawn in the figure).

(ii) Since the full curve $(1 \to 2 \to 3 \to 4 \to 1)$ is closed, this implies that the solutions to the harmonic oscillator equations of motion are periodic. In fact, it is easy to show that when (A, B) are arbitrary integration constants using the initial conditions, $x(0) = x_0$ and $y(0) = v_0 = y_0$, we obtain

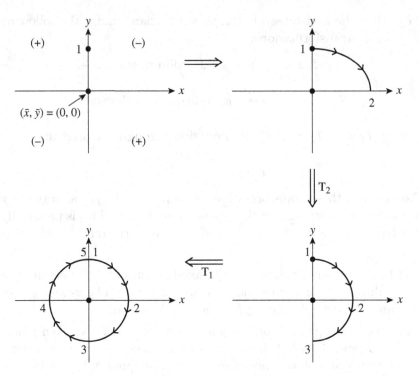

FIGURE 3.5 Steps for constructing a typical trajectory for the harmonic oscillator system. Equation (3.7.2) or (3.7.5).

$$x(t) = A \cos t + B \sin t, \quad y(t) = -A \sin t + B \cos t, \qquad (3.7.8)$$

$$x(t) = x_0 \cos t + y_0 \sin t, \quad y(t) = -x_0 \sin t + y_0 \cos t. \qquad (3.7.9)$$

(iii) Most importantly, our qualitative analysis has shown (geometrically) that all the solutions to the harmonic oscillator are bounded and periodic, and this was determined without knowledge of the analytic solutions to the equations of motion.

3.7.2 Damped Harmonic Oscillator

The differential equation modeling the damped harmonic oscillator is

$$\frac{d^2 x}{dt^2} + 2\epsilon \frac{dx}{dt} + x = 0; \quad x(0) = x_0, \quad \frac{dx(0)}{dt} = v_0. \qquad (3.7.10)$$

In system form, we have

$$\frac{dx}{dt} = y, \quad \frac{dy}{dt} = -x - 2\epsilon y; \quad x(0) = x_0, y(0) = v_0 = y_0. \quad (3.7.11)$$

There is a single fixed-point located at

$$(\bar{x}, \bar{y}) = (0, 0). \quad (3.7.12)$$

Also, for our purposes, assume that the 'damping coefficient', ϵ, is small and positive.

The differential equation determining the trajectories, $y = y(x)$, in phase-space is

$$\frac{dy}{dx} = -\left(\frac{x + 2\epsilon y}{y}\right), \quad (3.7.13)$$

and the two nullclines are (using the notation $y' \equiv dy/dx$)

$$\begin{cases} y' = 0 : y_0(x) = -\left(\frac{1}{2\epsilon}\right)x, \\ y' = \infty : y = 0 \text{ or the } x\text{-axis}. \end{cases} \quad (3.7.14)$$

Close inspection of the main features of the $x - y$ phase-plane for the damped harmonic oscillator, see Figure 3.6, allows the following conclusions to be made:

(a) There is only one fixed-point, $(\bar{x}, \bar{y}) = (0, 0)$.

(b) For $\epsilon > 0$, this fixed-point is a stable spiral.

(c) The general solution of the damped harmonic oscillator equations, see Equations (3.7.10) or (3.7.11), is expected to have the form sketched in Figure 3.7.

3.7.3 Nonlinear Cubic Oscillator

This oscillator has the equation of motion

$$\frac{d^2 x}{dt^2} + x^3 = 0, \, x(0) = x_0, \, \frac{dx(0)}{dt} = y_0. \quad (3.7.15)$$

The corresponding system equations are

$$\frac{dx}{dt} = y, \quad \frac{dy}{dt} = -x^3, \quad (3.7.16)$$

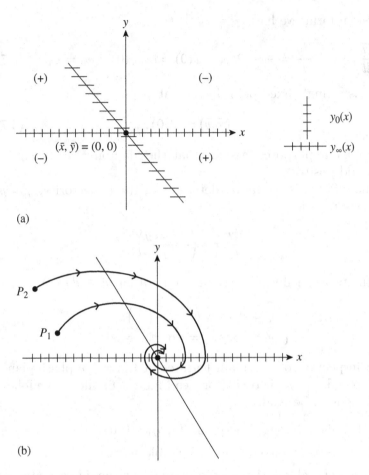

FIGURE 3.6 (a) Fixed-point, nullclines, and 'sign' of dy/dx in each of the four domains determined by the two nullclines, $y_0(x)$ and $y_\infty(x)$. Sketches of two trajectories for the damped harmonic oscillator.

with the same initial conditions as given previously.

This system has one fixed-point at $(\bar{x}, \bar{y}) = (0, 0)$. From the trajectory differential equation

$$\frac{dy}{dx} = -\frac{x^3}{y}, \qquad (3.7.17)$$

it is seen that Equation (3.7.17) has the same invariant properties as the harmonic oscillator; see the discussion after Equation (3.7.7).

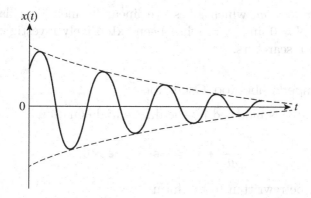

FIGURE 3.7   Plot of expected features of the solution to the damped harmonic oscillator equation.

Also, observe that a first-integral of Equation (3.7.17) is

$$\left(\frac{1}{2}\right)y^2 + \left(\frac{1}{4}\right)x^4 = \left(\frac{1}{2}\right)y_0^2 + \left(\frac{1}{4}\right)x_0^4 = \text{constant}. \tag{3.7.18}$$

Further, the nullclines for this nonlinear oscillator are exactly the same as the linear, harmonic oscillator; see Equation (3.7.6). However, there is a major difference in the nature of their fixed-points at $(\bar{x}, \bar{y}) = (0, 0)$. While both are 'centers', the harmonic oscillator is a 'linear center' and the cubic oscillator is a 'nonlinear center'. In terms of their respective phase-plane trajectories, the two sets of curves are topologically equivalent to each other. Hence, based on this discussion, it can be concluded that for each x_0 and y_0, such that $(x_0, y_0) \neq (0, 0)$, the cubic oscillator has periodic solutions. For this particular oscillator equation, the exact analytic solutions are known and may be expressed in terms of the Jacobi cosine and sine functions.

The arguments of this section can be extended to systems of the form

$$\begin{cases} \dfrac{d^2x}{dt^2} + x^{\frac{2N+1}{2m+1}} = 0, \\ N = (0, 1, 2, ...)\,,\, m = (0, 1, 2, ...) \end{cases} \tag{3.7.19}$$

Thus, this nonlinear differential equation has one fixed-point at $(\bar{x}, \bar{y})$, which is a nonlinear center and all solutions are periodic.

$$(\bar{x}, \bar{y}) = (0, 0). \tag{3.7.20}$$

Except for $n = m$, which gives the linear harmonic oscillator, the case with $N = 0$ and $m = 1$ has been extensively investigated by a number of researchers.

3.7.4 Damped Cube-Root Oscillator

This oscillator is modeled by the differential equation

$$\frac{d^2x}{dt^2} + x^{\frac{1}{3}} = -\epsilon\frac{dx}{dt}, \quad \epsilon > 0, \tag{3.7.21}$$

which can be rewritten to the form

$$\frac{dx}{dt} = y, \quad \frac{dy}{dt} = -\epsilon y - x^{\frac{1}{3}}. \tag{3.7.22}$$

From these latter equations, we determine that there is a single fixed-point at $(\bar{x}, \bar{y}) = (0, 0)$. Also, the trajectories in the $x - y$ phase-plane are solutions of the differential equation

$$\frac{dy}{dx} = -\left(\frac{x^{\frac{1}{3}} + \epsilon y}{y}\right), \tag{3.7.23}$$

and from this expression the two nullclines are found to be

$$\begin{cases} y' = 0: \text{ along the curve} y_0(x) = -\left(\frac{1}{\epsilon}\right) x^{\frac{1}{3}}, \\ y' = \infty: y_\infty \text{ is the } x\text{-axis.} \end{cases} \tag{3.7.24}$$

Figure 3.8a gives the major features of this phase-space, with a typical trajectory sketched in Figure 3.8b.

The qualitative, geometric analysis coming from the detailed examination of Figure 3.8 allows the following conclusions:

(i) Given an initial point $(x_0, y_0) \neq (0, 0)$, then its trajectory spirals into the fixed-point located at $(\bar{x}, \bar{y}) = (0, 0)$.

(ii) The fixed-point, $(\bar{x}, \bar{y}) = (0, 0)$, is stable.

(iii) The following 'sum rule' holds

$$\epsilon \int_0^\infty [y(t)]^2 \, dt = \left(\frac{1}{2}\right) y_0^2 + \left(\frac{3}{4}\right) x_0^{\frac{4}{3}}. \tag{3.7.25}$$

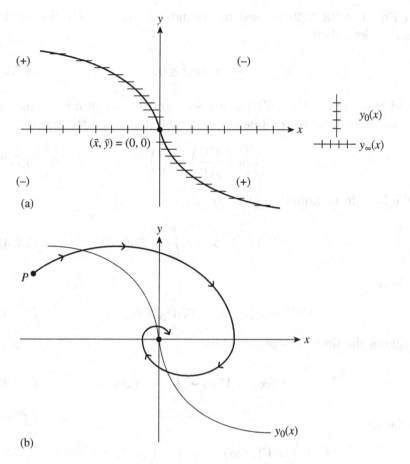

FIGURE 3.8 Phase-space for the damped cube-root oscillator given in Equation (3.7.22).

The arguments to support these results will now be given.

First, define $V(x, y)$ to be

$$V(x, y) = \left(\frac{1}{2}\right) y^2 + \left(\frac{3}{4}\right) x^{\frac{4}{3}}. \qquad (3.7.26)$$

Then,

$$V(0, 0) = 0 \quad ; \quad V(x, y) > 0, \text{ for } x \neq 0, y \neq 0. \qquad (3.7.27)$$

Further,

$$\frac{dV}{dt} = y\frac{dy}{dt} + x^{\frac{1}{3}}\frac{dx}{dt}, \qquad (3.7.28)$$

62 ■ Introduction to Qualitative Methods for Differential Equations

If Equation (3.7.22) is used to evaluate the right-hand side of this expression, then

$$\frac{dV}{dt} = -\epsilon y^2 \leq 0, \qquad (3.7.29)$$

and this implies that $V(t)$ is a non-negative, monotonic decreasing function of t. Since the phase-space analysis implies the result

$$\lim_{t \to \infty}\begin{pmatrix} x(t) \\ y(t) \end{pmatrix} = \begin{pmatrix} 0 \\ 0 \end{pmatrix}, \qquad (3.7.30)$$

We have from Equation (3.7.29)

$$V(t) = V(0) - \epsilon \int_0^t y(z)^2 \, dz, \qquad (3.7.31)$$

where

$$V(0) = V(x_0, y_0) > 0, \text{ if } x_0 \neq 0, y_0 \neq 0. \qquad (3.7.32)$$

Taking the limit, $t \to \infty$, we have

$$V(\infty) = V(0) - \int_0^\infty y(z)^2 \, dz, \qquad (3.7.33)$$

where

$$V_\infty = V[x(\infty), y(\infty)] = V(0,0) = 0. \qquad (3.7.34)$$

Substituting this last result into Equation (3.7.33) gives Equation (3.7.25).

3.7.5 $\ddot{x} + (1 + \dot{x})x = 0$

This differential equation models a system for which the angular frequency depends on the first derivative, i.e.,

$$\frac{d^2x}{dt^2} + \left(1 + \frac{dx}{dt}\right)x = 0, \qquad (3.7.35)$$

or

$$\frac{d^2x}{dt^2} + \Omega^2 x = 0, \quad \Omega^2 = \left(1 + \frac{dx}{dt}\right). \qquad (3.7.36)$$

Note that there are three possible solution behaviors; they are

$$\begin{cases} \frac{dx}{dt} < -1 & : \Omega^2 < 0, \text{unbounded solutions}, \\ \frac{dx}{dt} = -1 & : \Omega^2 = 0, \text{linear motion}, \\ \frac{dx}{dt} > -1 & : \Omega^2 > 0, \text{periodic solutions}. \end{cases} \quad (3.7.37)$$

Also,
$$x(t) = -t + x_0 \quad (3.7.38)$$
is a solution to Equation (3.7.35).

The system equations for Equation (3.7.35) are
$$\frac{dx}{dt} = y, \quad \frac{dy}{dt} = -(1+y)x \quad (3.7.39)$$

and there is one fixed-point at $(\bar{x}, \bar{y}) = (0, 0)$. Further, the two nullclines are
$$\begin{cases} y' = 0 & : \quad x = 0 \text{ or the } y\text{-axis, and } y = -1; \\ y' = \infty & : \quad x = 0 \text{ or the } x\text{-axis}, \end{cases} \quad (3.7.40)$$

since
$$\frac{dy}{dx} = -\frac{(1+y)x}{y}. \quad (3.7.41)$$

With regard to symmetry transformation, only reflection in the y-axis exists, i.e.,
$$T_1 : x \to -x, \quad y \to y. \quad (3.7.42)$$

The sketches of typical phase-space trajectories given in Figure 3.96 shows this.

Equation (3.7.41) is a separable differential equation and can be exactly integrated, i.e., it can be rewritten as
$$\int \frac{y}{1+y} dy + \int x \, dx = 0, \quad (3.7.43)$$

and carrying out the integrations gives the first-integral
$$I(x, y) = y - L_n(1+y) + \frac{x^2}{2} = \text{constant}. \quad (3.7.44)$$

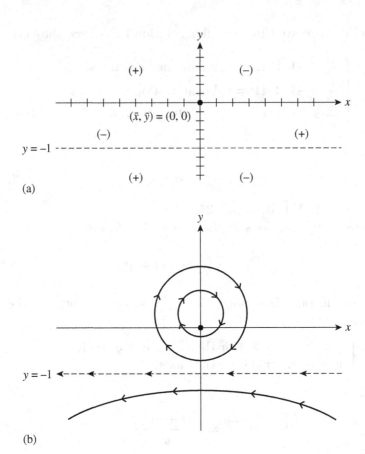

FIGURE 3.9 $x-y$ plane information for Equation (3.7.39).

Observe that for small enough x and y, i.e.,

$$|x| \ll 1, |y| \ll 1, \qquad (3.7.45)$$

we have

$$I(x, y) = \left(\frac{1}{2}\right) y^2 + \left(\frac{1}{2}\right) x^2 + O(y^3), \qquad (3.7.46)$$

where the first two terms on the right-hand side are the first-integral for the harmonic oscillators. This implies that the trajectories close to the fixed-point $(\bar{x}, \bar{y}) = (0, 0)$ are nearly circles. Finally, the inspection of Figure 3.96 further shows that the line, $y = -1$, separates the periodic and unbounded solutions.

3.7.6 Simple Predator–Prey Model

Consider two interacting populations: rabbits and foxes. A simple predator–prey model by making the following assumptions:

(a) If no foxes are present, the rabbits reproduce at a constant rate, k.
(b) If only foxes are present, they die at a rate proportional to the size of their total population.
(c) When both rabbits and foxes are present, the rabbits die at a rate proportional to the product of the two populations.
(d) Likewise, when both rabbits and foxes are present, the foxes increase at a rate proportional to the product of their two populations.

If we denote, respectively, the rabbit and fox populations by $x(t)$ and $y(t)$, then the above assumptions translate into the mathematical model:

$$\frac{dx}{dt} = k - r_1 xy, \quad \frac{dy}{dt} = r_2 xy - r_3 y, \qquad (3.7.47)$$

where (k, r_1, r_2, r_3) are positive parameters.

If we scale the variables in Equation (3.7.47) and set the remaining dimensionless parameters to one, then we have

$$\frac{dx}{dt} = 1 - xy, \quad \frac{dy}{dt} = xy - y. \qquad (3.7.48)$$

This set of equations has a fixed-point at $(\bar{x}, \bar{y}) = (1, 1)$ and from the trajectory determining differential equation

$$\frac{dy}{dx} = \frac{y(x-1)}{1-xy}, \qquad (3.7.49)$$

we can calculate the nullclines; they are

$$\begin{cases} y' = 0 : & y = 0 \text{ or the } x\text{-axis and } x = 1; \\ y' = \infty : & y_\infty(x) = \frac{1}{x}. \end{cases} \qquad (3.7.50)$$

With this information, we can draw a typical trajectory in the x–y plane (see Figure 3.10). We see that starting at an initial point in the first quadrant, then the trajectory that goes through that point spirals into the fixed-point at $(\bar{x}, \bar{y}) = (1, 1)$.

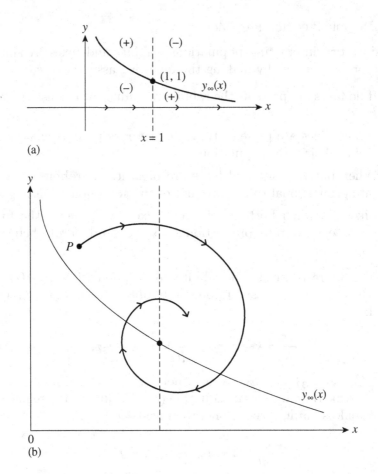

FIGURE 3.10 Phase-space of predator–prey model for Equation (3.7.48).

Another way of achieving this result is to do a linear stability analysis of the fixed-point. Let

$$x(t) = 1 + \alpha(t), \quad y(t) = 1 + \beta(t), \qquad (3.7.51)$$

where

$$|\alpha(0)| \ll 1, \quad |\beta(0)| \ll 1. \qquad (3.7.52)$$

Substituting these expressions into Equation (3.7.48) and keeping only linear terms give

$$\frac{d}{dt}\begin{pmatrix}\alpha\\\beta\end{pmatrix} = \begin{pmatrix}-1 & -1\\ 1 & 0\end{pmatrix}\begin{pmatrix}\alpha\\\beta\end{pmatrix}. \qquad (3.7.53)$$

The 2 × 2 matrix has the characteristic equation
$$\lambda^2 + \lambda + 1 = 0, \qquad (3.7.54)$$
which has the complex conjugate solutions
$$\lambda_1 = \lambda_2^* = -\left(\frac{1}{2}\right) + i\left(\frac{\sqrt{3}}{2}\right). \qquad (3.7.55)$$
The negative real parts of these roots imply that the fixed-point, $(\bar{x}, \bar{y}) = (1, 1)$, is a stable spiral point.

3.7.7 van der Pol Equation

The van der Pol equation provides a first approximation to many phenomena in the natural and physical sciences. Its differential equation is
$$\frac{d^2 x}{dt^2} + x = \epsilon(1 - x^2)\frac{dx}{dt}, 0 < \epsilon \ll 1. \qquad (3.7.56)$$
Variants of this equation have also being investigated, including the following forms
$$\frac{d^2 x}{dt^2} + x = \epsilon(1 - |x|)\frac{dx}{dt}, \qquad (3.7.57)$$
$$\frac{d^2 x}{dt^2} + x = \epsilon(1 - x^2)\left(\frac{dx}{dt}\right)^{\frac{1}{3}}, \qquad (3.7.58)$$
$$\frac{d^2 x}{dt^2} + x^{\frac{1}{3}} = \epsilon(1 - x^2)\frac{dx}{dt}. \qquad (3.7.59)$$
All of these four equations have solutions with the same qualitative properties for their solutions.

Equation (3.7.56) written in the system form is
$$\frac{dx}{dt} = y, \frac{dy}{dt} = -x + \epsilon(1 - x^2)y, \qquad (3.7.60)$$
where the parameter, ϵ, is taken to be positive. This system has a single fixed-point located at $(\bar{x}, \bar{y}) = (0, 0)$. Also, note that the linear approximation, which holds in a neighborhood of the fixed-point, is
$$\frac{d^2 x}{dt^2} + x = \epsilon\frac{dx}{dt}; |x| \ll 1, \ |y| \ll 1. \qquad (3.7.61)$$

68 ■ Introduction to Qualitative Methods for Differential Equations

Since the characteristic equation for Equation (3.7.41) is

$$\lambda^2 - \epsilon\lambda + 1 = 0, \qquad (3.7.62)$$

with solutions

$$\lambda_1 = \lambda_2^* = \left(\frac{1}{2}\right)\left[\epsilon + i\sqrt{4-\epsilon^2}\right]$$

$$\simeq \left(\frac{1}{2}\right)\epsilon + i, \qquad (3.7.63)$$

it follows that the fixed-point, $(\bar{x}, \bar{y}) = (0,0)$, of the van der Pol oscillator is an unstable spiral.

The trajectory determining differential equation is

$$\frac{dy}{dx} = \frac{-x + \epsilon(1-x^2)y}{y}. \qquad (3.7.64)$$

From this equation, we find the two nullclines to be

$$\begin{cases} y' = 0: \text{ along the curve } y_0(x) = \left(\frac{1}{\epsilon}\right)\left(\frac{x}{1-x^2}\right), \\ y^1 = \infty: y = 0 \text{ or along the } x\text{-axis}. \end{cases} \qquad (3.7.65)$$

We represent in Figure 3.11, the major features of the $x - y$ phase-plane for the van der Pol oscillator.

The next figure sketches a typical phase-space trajectory for initial conditions far from the fixed-point located at the origin. Note that for this case, the trajectory spirals inward toward the fixed-point. But, we know that trajectories in the neighborhood of the origin must spiral outward from it. So, how do we resolve this seemingly paradox, i.e., distant trajectories spiraling toward the origin, close to the origin spiraling from the origin? The resolution is that there must exist a simple closed curve around the fixed-point such that all trajectories spiral toward this closed curve (see Figure 3.13). This simple closed curve is called a limit cycle and represents a stable periodic solution of the van der Pol equation; in fact, it is the only periodic solution. For small ϵ, this solution is given by the approximation.

FigurerePhase-space trajectories for the van der Pol equation. The closed curve is the periodic solution. All other trajectories spiral in or out to this closed solution.

$$x_p(t) = 2\cos t + O(\epsilon). \qquad (3.7.66)$$

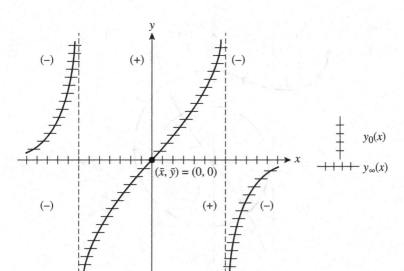

FIGURE 3.11 Phase-space features for the van der Pol equation.

In general, we have for any solution of the van der Pol differential equation the result (for $\epsilon \ll 1$)

$$X(t) \xrightarrow{\text{large } t} X_p(t). \qquad (3.7.67)$$

3.7.8 SIR Model for Disease Spread

The final system considered is a rather elementary model, which was constructed to provide insight into the spread of certain types of diseases. Since our interest is in the mathematical model itself rather than the details of its formulation, we will start with these equations. For readers interested in the formulation of models for the spread of disease, there is a vast literature on the subject and we provide a short list of such items in the Comments and Reference section.

The particular model we examine is

$$\frac{dS}{dt} = -\beta S\left(\frac{I}{N}\right), \qquad (3.7.68)$$

$$\frac{dI}{dt} = \beta S\left(\frac{I}{N}\right) - \gamma I, \qquad (3.7.69)$$

70 ■ Introduction to Qualitative Methods for Differential Equations

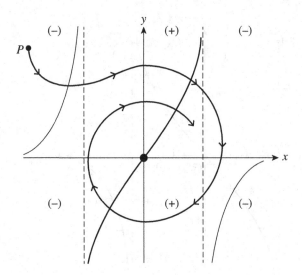

FIGURE 3.12 Trajectory for the van der Pol equation. starting at a point $P(x_0, y_0)$ where $|x_0| \gg 1$ and $|y_0| \gg 1$. All such trajectories spiral toward the fixed-point.

$$\frac{dR}{dt} = \gamma I, \tag{3.7.70}$$

$$N(t) = S(t) + I(t) + R(t) = \text{constant}, \tag{3.7.71}$$

where

- S(t) is the susceptible population, i.e., those members of the general population who can become infected, but are not.
- I(t) is the infected population, i.e., infected individuals who upon contact with susceptible individuals can infect susceptible persons.
- R(t) is the recovered population consisting of those individuals who have recovered from the disease.

We make the assumption that individuals in $R(t)$ have permanent immunity and never return to the susceptible population.

The above mathematical model allows for the prediction of how individuals transition from one population class to another, i.e.,

$$S(t) \rightarrow I(t) \longrightarrow R(t). \tag{3.7.72}$$

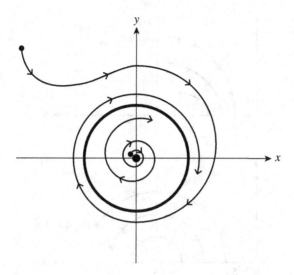

FIGURE 3.13 Phase-space trajectories for the van der Pol equation. The closed curve is the periodic solution. All other trajectories spiral in or out to this closed solution.

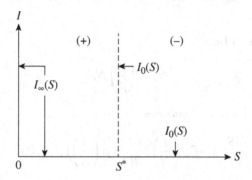

FIGURE 3.14 Phase-space for the SIR model using Equation (3.7.48).

Note that the mathematical structure of the model, Equations (3.7.68)–(3.7.70), imply the correctness of Equation (3.7.71), i.e., the whole population is constant. Also, since the variable $R(t)$ appears only in the third modeling equation, we can carry out our analysis with just the first two equations.

$$\frac{dS}{dt} = -\beta S\left(\frac{I}{N}\right), \frac{dI}{dt} = \beta S\left(\frac{I}{N}\right) - \gamma I. \qquad (3.7.73)$$

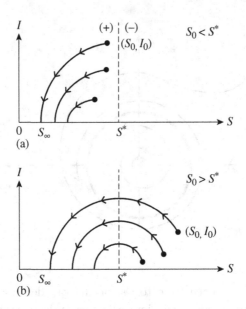

FIGURE 3.15 Two trajectories for the SIR disease spreading model. If $S_0 > S^*$, then an epidemic takes place. If $S_0 < S^*$, no epidemic occurs.

If S^* is defined to be

$$S^* = \left(\frac{\gamma}{\beta}\right) N, \qquad (3.7.74)$$

then Equation (3.7.73) becomes

$$\frac{dS}{dt} = -\beta S\left(\frac{I}{N}\right), \quad \frac{dI}{dt} = \beta\left(\frac{I}{N}\right)(S - S^*), \qquad (3.7.75)$$

and

$$\frac{dI}{dS} = -1 + \frac{S^*}{S}. \qquad (3.7.76)$$

Note that this is a separable differential equation, which can be easily solved to give the result

$$I(t) + S(t) = I_0 + S_0 + S^* \mathrm{Ln}\left(\frac{S(t)}{S_0}\right), \qquad (3.7.77)$$

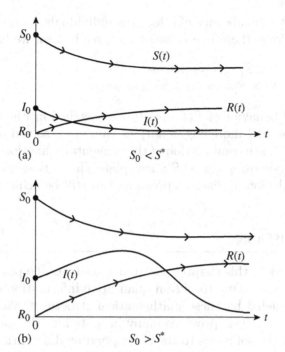

FIGURE 3.16 General time dependence of $S(t), I(t)$, and $R(t)$ for an SIR model. We assume $S_0 > 0, I_0 > 0$, and $R_0 = 0$.

Figure 3.14 gives the major geometric features of the I–S plane. The nullclines are determined from

$$\frac{dI}{ds} = \frac{dI/dt}{ds/dt} = \frac{\beta\left(\frac{I}{N}\right)(s - s^*)}{\beta s\left(\frac{I}{N}\right)}, \qquad (3.7.78)$$

from which we conclude ($I' = dI/dS$)

$I' = 0$: $I_0(x)$ consists of the S-axis and the
line $S = S^*$.

$I' = \infty$: $I_\infty(x)$ consists of the S- and I-axes.

An interesting aspect of this SIR model is that in the I–S plane, the S-axis is both $I_0(S)$ and $I_\infty(S)$ nullclines. Since points where these two types of nullclines intersect are fixed-points, it follows that we have a continuum of fixed-points and they lie along the (non-negative) S-axis. This makes sense because points on the S-axis correspond to the absence of infectious individuals. Consequently, in these situations, there is no disease to spread. Also, we have that

$$\begin{cases} 0 < S_0 < S^* \Rightarrow \text{disease dies out;} \\ S_0 > S^* \Longrightarrow \text{an epidemic occurs.} \end{cases} \quad (3.7.79)$$

The general behavior of $S(t)$ and $I(t)$ is sketched in Figure 3.16.

While the SIR modeling equations cannot be solved analytically in terms of a finite combination of the elementary functions, our qualitative analysis using the I–S phase-plane shows that great insights into the evolution of disease spreading can still be achieved.

3.8 DISCUSSION

The main task of this chapter was to demonstrate the power of using the 2-dim phase-plane to obtain qualitative information on systems that are modeled by these mathematical structures. We have seen that such an analysis provides many insights into the nature of the behaviors of the solutions to the corresponding differential equations even if no exact analytical solutions exist. This knowledge can also be used as an aid for the construction of 'valid' approximations to the actual solutions. Further, the understandings derived from a phase-plane diagram may provide information on constructing improved mathematical models for the system being examined.

Another important feature is that, at least for 2-dim systems, almost all of the 'work' can be done 'by hand'. The only requirement is the ability to be able to make reasonably accurate sketches of various types of curves.

Finally, it is clear that extending this methodology to dim −3 or higher systems will generally require the use of computers.

PROBLEMS

Instead of a list of problems, it is strongly recommended that the following books be consulted. They all contain many examples of 2-dim systems and their analysis using phase-plane techniques.

1. D. Basmadjian, *Mathematical Modeling of Physical Systems* (Oxford University Press, Oxford, 2003).

2. L. Edelstein-Keshet, *Mathematical Models in Biology* (McGraw-Hill, New York, 1987).

3. J. H. Lin, *A First Course in the Qualitative Theory of Differential Equations* (Prentice Hall, Upper Saddle River, NJ, 2003).

4. R. E. Mickens, *Mathematical Methods for the Natural and Engineering Sciences*, 2nd Edition (World Scientific, Singapore, 2017).

5. J. D. Murray, *Mathematical Biology* (Springer, New York, 1989).

6. A. Panfilov, Qualitative analysis of differential equations; arXiv:1803.05291v1 [math.GM] (2018, 117 pps.).

COMMENTS AND REFERENCES

This chapter is heavily based on Chapter 4 of my book listed in reference 4 above.

CHAPTER 4

Sturm–Liouville Problems

4.1 INTRODUCTION

Many physical systems can be modeled in one-space dimension by ordinary differential equations having the form

$$\frac{d^2x}{dt^2} = F\left(x, \frac{dx}{dt}\right). \qquad (4.1.1)$$

This particular structure is a consequence of the fact that these systems have dynamics that follow from the basic Newton's law of force, i.e.,

$$(\text{mass})\,(\text{acceleration}) = (\text{total})\,\text{force}. \qquad (4.1.2)$$

Similarly, many continuous physical phenomena have mathematical models that are either first- or second order in the time-independent variable. The purpose of this chapter is to investigate the behavior of the solutions to ordinary differential equations, which can be expressed as

$$a_0(x)\,y''(x) + a_1(x)\,y'(x) + a_2(x)\,y(x) = 0, \qquad (4.1.3)$$

considered as boundary-valued problems over the interval

$$a \leq x \leq b, \qquad (4.1.4)$$

where (a, b) may be either finite or unbounded. For this chapter, the following notation is used

$$y = y(x), \quad y' = \frac{dy(x)}{dx}, y'' = \frac{d^2y(x)}{dx^2}, \quad \text{etc.} \qquad (4.1.5)$$

We begin by showing that generally the first-derivative term in Equation (4.1.3) can be eliminated by a suitable change of the

Sturm–Liouville Problems ■ 77

dependent variable. Another interesting and valuable transformation is the one associated with the names of Liouville and Green, called appropriately the Liouville–Green transformation. This procedure is especially useful for calculating the asymptotic behaviors of solutions to linear, second-order differential equations.

4.1.1 Elimination of First-Derivative Term

Assume that the function $a_0(x)$ has no zeros in the interval given by Equation (4.1.3). If we divide this equation by $a_0(x)$ and make the definitions

$$p(x) = \frac{a_1(x)}{a_0(x)}, \quad q(x) = \frac{a_2(x)}{a_0(x)}, \qquad (4.1.6)$$

then we have

$$y''(x) + p(x) y'(x) + q(x) y(x). \qquad (4.1.7)$$

Let $I(x)$ be

$$I(x) = \left(\frac{1}{2}\right) \int^{(x)} p(z)\, dz, \qquad (4.1.8)$$

and make the following dependent variable transformation

$$y(x) = u(x)\, e^{-I(x)}. \qquad (4.1.9)$$

Then, we have

$$I'(x) = \frac{P(x)}{2}, \quad I''(x) = \left(\frac{1}{2}\right) P'(x), \qquad (4.1.10)$$

and

$$\begin{cases} y'(x) = [u' - I'u]\, e^{-I}, \\ y''(x) = \left\{u'' - 2I'u' + \left[(I')^2 - I''\right]u\right\} e^{-I}. \end{cases} \qquad (4.1.11)$$

Substitution of all of these results into Equation (4.1.7) gives, after some algebraic manipulation, the result

$$u''(x) + Q(x)\, u(x) = 0, \qquad (4.1.12)$$

where

$$Q(x) = q(x) - \left(\frac{1}{2}\right) p'(x) - \frac{p(x)^2}{4}. \qquad (4.1.13)$$

To illustrate this procedure, consider the differential equation

$$x^2 y'' + xy' + (x^2 - \lambda) y = 0, \qquad (4.1.14)$$

or upon division by x^2,

$$y'' + \left(\frac{1}{x}\right) y' + \left(1 - \frac{\lambda}{x^2}\right) y = 0. \qquad (4.1.15)$$

Therefore,

$$p(x) = \frac{1}{x}, \quad q(x) = 1 - \frac{\lambda}{x^2}, \qquad (4.1.16)$$

and

$$I(x) = \left(\frac{1}{2}\right) \int^x p(z)\, dz$$
$$= \left(\frac{1}{2}\right) \int^x \frac{dz}{z}$$
$$= \left(\frac{1}{2}\right) \operatorname{Ln}(x) = \operatorname{Ln}\left(\sqrt{x}\right), \qquad (4.1.17)$$

with

$$e^{-I(x)} = \frac{1}{\sqrt{x}}. \qquad (4.1.18)$$

The final result is that $u(x)$ satisfies the following differential equation:

$$u''(x) + \left[1 + \left(\frac{1 - 4\lambda}{4x^2}\right)\right] u(x) = 0. \qquad (4.1.19)$$

4.1.2 Liouville–Green Transformation

In the previous section, we showed that the equation

$$\frac{d^2 y(x)}{dx^2} + p(x) \frac{dy(x)}{dx} + q(x) y(x) = 0 \qquad (4.1.20)$$

could be transformed into

$$\frac{d^2 v(x)}{dx^2} + Q(x) v(x) = 0, \qquad (4.1.21)$$

where

$$V(x) = y(x) \exp\left[\left(\frac{1}{2}\right)\int^x p(z)\,dz\right] \quad (4.1.22)$$

and

$$Q(x) = q(x) - \left(\frac{1}{2}\right)p'(x) - \left(\frac{1}{4}\right)p(x)^2. \quad (4.1.23)$$

Let us now introduce a new independent variable, s, which is a function of x, i.e., $s = s(x)$ and it is defined by the relation

$$s = s(x) = \int^x \sqrt{Q(z)}\,dz. \quad (4.1.24)$$

Our goal is to replace the variable x in Equation (4.1.21) by new variable s. From the calculus, it follows that

$$\frac{dv}{ds} = \frac{dv}{ds}\frac{ds}{dx} = \sqrt{Q(x)}\frac{dv}{ds}. \quad (4.1.25)$$

Since from Equation (4.1.24)

$$\frac{ds}{dx} = \sqrt{Q(x)}. \quad (4.1.26)$$

Also,

$$\frac{d^2v}{dx^2} = Q(x)\frac{d^2v}{ds^2} + \left(\frac{1}{2\sqrt{Q(x)}}\right)\left(\frac{dQ(x)}{ds}\right)\frac{dv}{ds}. \quad (4.1.27)$$

The result of Equation (4.1.24) can be inverted to give x as a function of s, i.e., $x = x(s)$, and consequently, $Q(x) = Q(x(s))$. If we make these replacements in Equation (4.1.21), then we get the expression

$$\frac{d^2v}{ds^2} + \left[\left(\frac{1}{2}\right)\left(\frac{dQ}{dx}\right)\left(\frac{1}{Q^{3/2}}\right)\right]\frac{dv}{ds} + v = 0. \quad (4.1.28)$$

The first-derivative term can be eliminated by the following change of variable, i.e., $v(s) \to w(s)$, where

$$w(s) = v(s)\exp\left[\left(\frac{1}{2}\right)\int^s \left(\frac{dQ}{dz}\right)\left(\frac{1}{2Q^{3/2}}\right)dz\right]. \quad (4.1.29)$$

Since
$$ds = \frac{ds}{dx}dx = \sqrt{Q(x)}\,dx, \qquad (4.1.30)$$

Equation (4.1.29) can be rewritten as
$$w(s) = v(s)\left[\exp\left(\frac{1}{4}\right)\int^x \left(\frac{dQ}{dz}\right)\left(\frac{1}{Q}\right)dz\right]$$
$$= [Q]^{\frac{1}{4}} v(s). \qquad (4.1.31)$$

With this result, Equation (4.1.28) becomes
$$\frac{d^2w}{ds^2} + \left[1 - \left(\frac{1}{2}\right)\frac{dh}{ds} - \left(\frac{1}{4}\right)h^2\right]w = 0, \qquad (4.1.32)$$

where
$$h(s) = \left(\frac{1}{2}\right)\left(\frac{dQ}{dx}\right)\left(\frac{1}{Q^{3/2}}\right). \qquad (4.1.33)$$

This series of variable transformations, i.e.,
$$y(x) \longrightarrow v(x) \longrightarrow v(s) \longrightarrow w(s), \qquad (4.1.34)$$

is the Liouville–Green transformation and will be used later in this chapter to calculate the behavior of solutions to certain differential equations for large x.

4.2 THE VIBRATING STRING

As a preliminary introduction to Sturm–Liouville problems, we examine in some detail a vibrating finite-length string of length L. Let it be aligned with the x-axis. The equation of motion is
$$\frac{d^2u(x)}{dx^2} + \lambda u(x) = 0, \qquad (4.2.1)$$

where x denotes the location of a point on the string and $u(x)$ is the distance of that point from the x-axis (see Figure 4.1).

The parameter λ is a priori unknown and, as we will show, is determined by the values of $u(x)$ and $du(x)/dx$ selected at $x = 0$ and $x = L$. The four possibilities are

$$(1)\quad x(0) = 0, \quad x(L) = 0; \qquad (4.2.2)$$

FIGURE 4.1 String along the x-axis, $0 \le x \le L$, $u(x)$ is the displacement from the x-axis.

$$(2) \quad x(0) = 0, \quad x'(L) = 0; \tag{4.2.3}$$

$$(3) \quad x'(0) = 0, \quad x(L) = 0; \tag{4.2.4}$$

$$(4) \quad x'(0) = 0, \quad x'(L) = 0. \tag{4.2.5}$$

Note that cases (2) and (3) are mathematically equivalent and we will only consider (2). Physically, the four cases correspond to the following situations:

(1) The string is clamped at both ends;
(2) The string is clamped at $x = 0$, but free at $x = L$;
(3) The string is free at $x = 0$, but clamped at $x = L$;
(4) Both ends are free to move.

Also note that since we wish to have periodic solutions, we assume $\lambda > 0$.

4.2.1 Both Ends Fixed

The general solution to Equation (4.2.1) is

$$u(x) = A \sin\left(\sqrt{\lambda}\, x\right) + B \cos\left(\sqrt{\lambda}\, x\right). \tag{4.2.6}$$

Since $u(0) = 0$ and $u(L) = 0$, we have

$$\begin{cases} u(0) = 0 \Rightarrow B = 0; \\ u(L) = 0 \Rightarrow A \sin\left(\sqrt{\lambda}\, L\right) = 0, \end{cases} \tag{4.2.7}$$

or

$$\sqrt{\lambda}\, L = n\pi, \quad n = (1, 2, 3, \ldots). \tag{4.2.8}$$

82 ■ Introduction to Qualitative Methods for Differential Equations

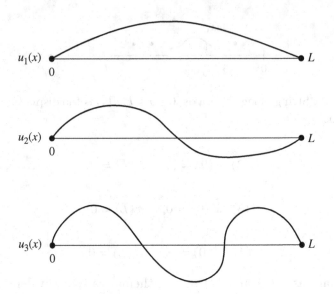

FIGURE 4.2 First three modes of vibration for the case $u(0) = 0$ and $u(L) = 0$.

However, the length of the string is constant, consequently, it follows that the values of λ can only take on the discrete values

$$\lambda_n = \left(\frac{n\pi}{L}\right)^2. \tag{4.2.9}$$

Therefore, the solutions of Equation (4.2.1) for the boundary conditions, $u(0) = 0$ and $u(L) = 0$, are

$$U_n(x) = A_n \sin\left(\frac{n\pi x}{L}\right). \tag{4.2.10}$$

Since the differential equation is linear, we do not have enough information to determine the A_n. Figure 4.2 gives sketches of the first three modes of vibrations.

4.2.2 One Fixed and One Free Ends

For this case, the general solution is still Equation (4.2.6), but the initial conditions are now those given in Equation (4.2.3). Therefore,

$$\begin{cases} u(0) = 0 \Rightarrow B = 0; \\ U'(L) = 0 \Rightarrow u'L = -\sqrt{\lambda}\cos\left(\sqrt{\lambda}\,L\right) = 0. \end{cases} \tag{4.2.11}$$

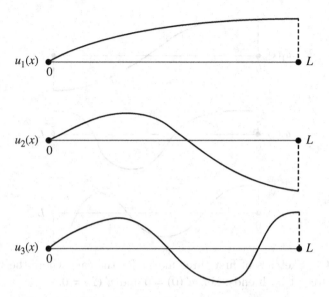

FIGURE 4.3 First three modes of a vibrating string in which one end is fixed and the other is free, i.e., $u(0) = 0$ and $u'(L) = 0$.

But,

$$\cos\left(\sqrt{\lambda}\, L\right) = 0 \Rightarrow \sqrt{\lambda}\, L = (2n-1)\left(\frac{\pi}{2}\right), \quad n = (1, 2, 3, \ldots), \quad (4.2.12)$$

and therefore we have for λ the following set of discrete values

$$\lambda_n = \left[\frac{(2n-1)\pi}{2L}\right]^2, \quad n = (1, 2, 3, \ldots). \quad (4.2.13)$$

Therefore, the associated solutions, $u_n(x)$, are

$$u_n(x) = A_n \sin\left[\frac{(2n-1)\pi x}{2L}\right]. \quad (4.2.14)$$

Figure 4.3 sketches the general shapes of the first three modes for $u(0) = 0$ and $u'(L) = 0$.

4.2.3 Both Ends Free

The general solution is Equation (4.2.6), it follows that

$$u'(x) = \sqrt{\lambda}\, \cos\left(\sqrt{\lambda}\, x\right) - \sqrt{\lambda}\, B \sin\left(\sqrt{\lambda}\, x\right), \quad (4.2.15)$$

84 ■ Introduction to Qualitative Methods for Differential Equations

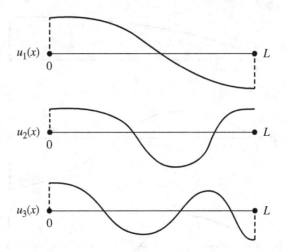

FIGURE 4.4 Sketches of first three modes for the care where the vibrating string is free at both ends, i.e., $u'(0) = 0$ and $u'(L) = 0$.

and

$$\begin{cases} u'(0) = 0 \Longrightarrow A = 0; \\ u'(L) = 0 \Longrightarrow B \sin\left(\sqrt{\lambda} L\right) = 0. \end{cases} \quad (4.2.16)$$

and, therefore,

$$\lambda_n = \left(\frac{n\pi}{L}\right)^2, \quad (4.2.17)$$

with

$$u_n(x) = B_n \cos\left(\frac{n\pi x}{L}\right), \quad n = (1, 2, ...). \quad (4.2.18)$$

The modes $u_1(x)$, $u_2(x)$ and $u_3(x)$ are sketched in Figure 4.4.

4.2.4 Summary

Our study of the vibratory modes of a finite length allows the following conclusions to be reached:

(a) The type of boundary conditions selected determines the values of λ and its associated function $u(x, \lambda)$.

(b) For a given set of boundary conditions, λ takes on a set of unbounded discrete values.

(c) There is a smallest value for λ and they can be ordered in an increasing unbounded sequence, i.e., with

$$\lambda_1 < \lambda_2 < \lambda_3 < \cdots < \lambda_n < \cdots, \qquad (4.2.19)$$

$$\underset{n \to \infty}{Lim} \lambda_n = \infty. \qquad (4.2.20)$$

(d) For a given set of boundary conditions and for a particular λ_n, the functions $u_n(x)$ have a definite number of zeros whose value depends on n.

(e) Note that the boundary conditions, $u(0) = 0$ and $u(L) = 0$, and $u'(0) = 0$ and $u'(L) = 0$, have the same set of λ values, and the characteristic shape functions differ.

From a physics perspective, our analysis of the finite length string shows that the different ways of clamping the string give rise to entirely different modes of oscillation.

We now turn to a major generalization of the results of this section, namely, the 'Sturm–Liouville problem'. However, before doing so, we state several results related to the comparison of solutions to linear, second-order, ordinary differential equations. These are presented in the next section.

4.3 SEPARATION AND COMPARISON RESULTS

We now state, without proofs, several interesting and important results regarding the solutions to the linear, second-order differential equation

$$y''(x) + a_1(x) y'(x) + a_2(x) y(x) = 0. \qquad (4.3.1)$$

We assume that $a_1(x)$ and $a_2(x)$ are defined and continuous on the interval, $a \leq x \leq b$. This equation has two linearly independent solutions, $Y_1(x)$ and $Y_2(x)$, and, consequently, its general solution is

$$y(x) = c_1 y_1(x) + c_2 y_2(x), \qquad (4.3.2)$$

where c_1 and c_2 are arbitrary constants.

86 ■ Introduction to Qualitative Methods for Differential Equations

The following theorem provides information on the interconnections of the zeros of $y_1(x)$ and $y_2(x)$.

Theorem 1 Let $y_1(x)$ and $y_2(x)$ be linearly independent solutions of Equation (4.3.1), then $y_1(x)$ must have a zero between any two consecutive zeros of $y_2(x)$ and, likewise, $y_2(x)$ must have a zero between any two consecutive zeros of $y_1(x)$.

Another way of stating this result is that the zeros of $y_1(x)$ and $y_2(x)$ alternate. This powerful result is called the *Sturm separation theorem*. An elementary example involves the sine and cosine functions, which are solutions to the differential equation

$$\frac{d^2 x}{d\theta^2} + x = 0 \implies x(\theta) = A\cos\theta + B\sin\theta, \qquad (4.3.3)$$

and where

$$\begin{cases} \sin(0) = 0 & \cos\left(\frac{\pi}{2}\right) = 0 \\ \sin(\pi) = 0 & \cos\left(\frac{3\pi}{4}\right) = 0 \end{cases}. \qquad (4.3.4)$$

The next theorem allows knowledge of the solution behaviors of one differential equation to restrict the solution behaviors of a related second differential equation if certain conditions hold.

Theorem 2 Let $p(x)$ and $q(x)$ be defined on the interval, $a \leq x \leq b$, and let them have the property

$$p(x) \geq q(x), \qquad a \leq x \leq b. \qquad (4.3.5)$$

Let $f(x)$ and $g(x)$ be, respectively, solutions of

$$u''(x) + p(x)u(x) = 0 \quad \text{and} \quad v''(x) + q(x)v(x) = 0. \qquad (4.3.6)$$

Then $f(x)$ has at least one zero between any two zeros of $g(x)$, unless $p(x) \equiv q(x)$ and $f(x)$ and $g(x)$ are linearly dependent.

An interesting consequence of this theorem is that if $q(x) \leq 0$, then no non-trivial solution of

$$V''(x) + q(x)v(x) = 0, \qquad q(x) \leq 0, \qquad (4.3.7)$$

Can have more than one zero.

Another result is that if $q(x) = \lambda^2 > 0$, where λ is a constant, and if $p(x) \geq q(x) = \lambda^2$, then every solution of

$$u''(x) + p(x)u(x) = 0, \quad p(x) \geq \lambda^2 > 0, \quad (4.3.8)$$

must have a zero between any two consecutive zeros of the solutions to

$$v''(x) + \lambda^2 v(x) = 0 \quad (4.3.9)$$

Note that since the general solution of Equation (4.3.9) is

$$v(x) = A \sin[\lambda(x - x_0)], \quad (4.3.10)$$

then under these conditions, $u(x)$ must have at least one zero in every interval of x of length (π/λ).

A corollary to Theorem 2 is that if

$$y''(x) + p(x)y(x) = 0 \quad (4.3.11)$$

has all nontrivial solutions oscillatory, then if $P(x) \geq p(x)$, it follows that all solutions of

$$u''(x) + P(x)u(x) = 0 \quad (4.3.12)$$

are oscillatory.

4.3.1 $y'(x) + f(x)y(x) = 0$

The differential equation

$$y''(x) + f(x)y(x) = 0 \quad (4.3.13)$$

appears in many areas of the science. In particular, the time-independent Schrodinger equation (TISE) in one space dimension takes this form. Based on the discussion of the previous section, the following conclusions can be reached:

(i) If there exists an interval, $x_1 < x < x_2$, such that $f(x)$ is positive on this interval, then $y(x)$ is oscillatory.

(ii) If there exists an interval, $x_3 < x < x_4$, such that $f(x)$ is negative, then $y(x)$ has two 'exponential type' components, one increasing and the other decreasing.

To illustrate these results consider the Airy differential equation

$$y''(x) + xy(x) = 0, \quad -\infty < x < +\infty, \quad (4.3.14)$$

where $f(x) = x$. Since

$$f(x) = x : \begin{cases} > 0, \text{ for } x > 0; \\ < 0, \text{ for } x < 0, \end{cases} \quad (4.3.15)$$

it follows that $y(x)$ is oscillatory for $x > 0$ and 'exponentially' increasing/decreasing for $x < 0$. Because the Airy equation is a linear, second-order differential equation, there are two linearly independent solutions. For $x > 0$, we write the solution as

$$x > 0 : \quad y_+(x) = C(x) + S(x), \quad (4.3.16)$$

where $C(x)$ and $S(x)$ are bounded, oscillatory functions similar in form to the trigonometric cosine and sine functions. Note that $C(x)$ and $S(x)$ are oscillatory, but not in general periodic.

For $x < 0$, the solution takes the form

$$x < 0 : \quad y_-(x) = E_+(x) + E_-(x), \quad (4.3.17)$$

where $E_+(x)$ and $E_-(x)$ are, respectively, exponential-type increasing and decreasing functions.

Figure 4.5 provides sketches of generic curves corresponding to $C(x), S(x), E_+(x)$ and $E_-(x)$.

The two linear independent solutions to the Airy equation are generally denoted by $A_i(x)$ and $B_i(x)$. In terms of the (for now unknown) functions given in Equations (4.3.16) and (4.3.17), these two functions are defined as follows

$$Ai(x) = \begin{cases} E_-(x), \text{ for } x < 0; \\ \text{Linear combination of } C(x) \\ \text{and } S(x), \text{ for } x > 0, \end{cases} \quad (4.3.18)$$

and

$$Bi(x) = \begin{cases} E_+(x), \text{ for } x < 0; \\ \text{Linear combination of} \\ C(x) \text{ and } S(x) \text{ for } x > 0. \end{cases} \quad (4.3.19)$$

Note that at $x = 0$, we require that the functions to the left and right match with regard to being continuous and having the same value

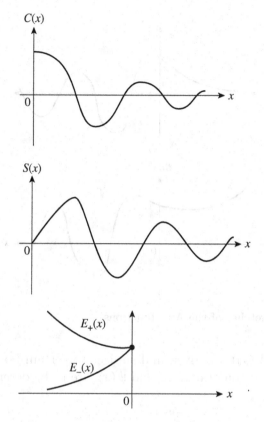

FIGURE 4.5 Sketches of generic plots of $C(x)$, $S(x)$, $E_+(x)$ and $E_-(x)$.

of the derivative. Based on these requirements, Figure 4.6 provides sketches of $Ai(x)$ and $Bi(x)$.

Finally, it should be remarked that from the qualitative results of the previous section, we have been able to derive a large number of the general features of the Airy functions. It should be clear that this methodology can be applied to any other differential equation that has the form of Equation (4.3.13). An important example is the time-independent Schrodinger equation:

$$-\left(\frac{\hbar^2}{2m}\right)\psi''(x) + V(x)\psi(x) = E\psi(x), \qquad (4.3.20)$$

which can be rewritten as

$$\psi''(x) + \left(\frac{2m}{\hbar^2}\right)[E - V(x)]\psi(x) = 0, \qquad (4.3.21)$$

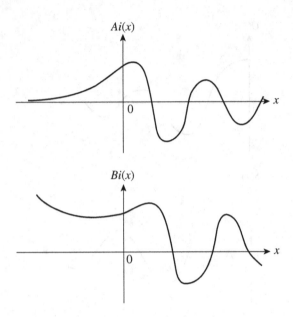

FIGURE 4.6 Sketches of the Airy functions.

where $(m, \hbar, V(x))$ are given and subject to certain bounded conditions and other constraints, E and $\psi(x)$, are to be determined.

4.4 STURM–LIOUVILLE PROBLEMS

Definition 1
Consider three real functions, $[p(x), q(x), T(x)]$, defined on the interval, $a \leq x \leq b$. Assume that they have the following properties:

(i) $p(x)$ has a continuous first derivative and has the additional property that $p(x) > 0$.

(ii) $q(x)$ and $r(x)$ are continuous, with $r(x) > 0$.

Let λ be a parameter that does not depend on x. Let real constants $[A_1, A_2, B_1, B_2]$ exist such that

$$\begin{cases} A_1 y(a) + A_2 y'(a) = 0, \\ B_1 y(b) + B_2 y'(b) = 0, \end{cases} \quad (4.4.1)$$

where (A_1, A_2) and (B_1, B_2) are not both equal to zero.

The second-order, linear differential equation

$$\frac{d}{dx}\left[p(x)\frac{dy}{dx}\right] + [q(x) + \lambda r(x)]\, y = 0, \qquad (4.4.2)$$

along with the boundary-value conditions of Equation (4.4.1) defines the *Sturm–Liouville problem*.

Definition 2
The values of the parameter λ for which the Sturm–Liouville problem has nontrivial solutions are called the *eigenvalues* of the problem. The associated solutions are called the *eigenfunctions* of the problem.

4.4.1 Properties of the Eigenvalues and Eigenfunctions

The following is a listing of some of the important properties of the eigenvalues and eigenfunctions for the Sturm–Liouville problem.

(1) For a particular Sturm–Liouville problem there are an infinite number of eigenvalues, all real, and they form an unbounded monotonic increasing sequence, i.e., with

$$\lambda_1 < \lambda_2 < \lambda_3 < \cdots < \lambda_n < \cdots, \qquad (4.4.3)$$

$$\underset{n\to\infty}{Lim}\lambda_n = \infty. \qquad (4.4.4)$$

(2) Associated with each eigenvalue, λ_n, there exists a unique eigenfunction, $\varphi_n(x)$, defined up to an overall multiplicative constant.

(3) In general, the eigenfunction $\varphi_n(x)$, associated with eigenvalue λ_n has exactly $(n-1)$ simple zeros in the open interval, $a < x < b$.

4.4.2 Orthogonality of Eigenfunctions

Definition 3
Let $f(x)$ and $g(x)$ be defined on the interval, $a \leq x \leq b$. Let there exist a function, $w(x) > 0$, defined on this interval, such that

$$\int_a^b f(x)\, g(x)\, w(x)\, dx = 0. \qquad (4.4.5)$$

Then $f(x)$ and $g(x)$ are said to be *orthogonal* on this interval with respect to the *weight function* $w(x)$.

Definition 4
Let $\{\psi_n(x) : n = 1, 2, 3, ...\}$ be an infinite set of functions defined on the interval, $a \leq x \leq b$.

Let
$$\int_a^b \psi_n(x) \psi_m(x) w(x) \, dx = 0, \, n \neq m. \tag{4.4.6}$$

Then the set $\{\psi_n(x) : n = 1, 2, 3, ...\}$ is called an orthogonal system on the interval, $a \leq x \leq b$, with respect to the weight function $w(x)$.

Theorem 3 Consider the Sturm–Liouville problem with its infinite set of corresponding eigenvalues and functions, i.e., $\{\lambda_n, \varphi_n(x) : n = 1, 2, 3, ...\}$. Then
$$\int_a^b \varphi_m(x) \varphi_n(x) r(x) \, dx = 0, \quad m \neq n. \tag{4.4.7}$$

This means that the eigenfunctions $\varphi_m(x)$ and $\varphi_n(x)$ are orthogonal with respect to the weight function $w(x) = r(x)$ on the interval, $a \leq x \leq b$.

4.4.3 Expansion of Functions

Definition 5
Let $f(x)$ be defined on the interval, $a \leq x \leq b$, such that the following integral exists
$$\int_a^b [f(x)]^2 w(x) \, dx = 1. \tag{4.4.8}$$

Then $f(x)$ is said to be *normalized* with respect to the weight function $w(x)$ on this interval.

Definition 6
Let $\{\psi_n(x) : n = 1, 2, 3, ...\}$ be an infinite set of functions defined on the interval, $a \leq x \leq b$.

Let
$$\int_a^b \psi_n(x) \psi_m(x) w(x) \, dx = \delta_{mn}, \tag{4.4.9}$$

where discrete delta function, also known as the Kronecker delta function, is defined as

$$\delta_{nm} = \delta_{mn} = \begin{cases} 0, & \text{if } m \neq n; \\ 1, & \text{if } m = n. \end{cases}$$

Theorem 4 Consider a Sturm–Liouville problem with its infinite set of eigenvalues and eigenfunctions, phi $\{\lambda_n, \varphi_n(x) : n = 1, 2, 3, ...\}$. Assume that $f(x)$ is continuous on the interval, $a \leq x \leq b$, and has a piecewise-continuous first-derivative on this interval. Further, assume that $f(x)$ and all the $\varphi_n(x)$ satisfy the same boundary conditions at $x = a$ and $x = b$. Then the series

$$\sum_{n=1}^{\infty} c_n \varphi_n(x) \tag{4.4.10}$$

where

$$c_n = \int_a^b f(x) \varphi_n(x) r(x) \, dx, \quad n = (1, 2, 3, ...), \tag{4.4.11}$$

converges uniformly and absolutely to $f(x)$ in the interval, $a \leq x \leq b$. In other words, under these conditions

$$f(x) = \sum_{n=1}^{\infty} c_n \varphi_n(x). \tag{4.4.12}$$

4.5 RELATED ISSUES

4.5.1 Reduction to Sturm–Liouville Form

We now show how to reduce the following second-order, linear differential equation to the Sturm–Liouville form,

$$a_0(x) y''(x) + a_1(x) y'(x) + a_2(x) y(x) = 0. \tag{4.5.1}$$

It can be directly verified that the following is an integrating factor for Equation (4.5.1),

$$\mu(x) = \left[\frac{1}{a_0(x)}\right] \exp\left[\int \frac{a_1(x)}{a_0(x)} \cdot dx\right]. \tag{4.5.2}$$

Therefore, multiplying Equation (4.5.1) by $\mu(x)$, the resulting expression can be written as

$$\frac{d}{dx}\left[\mu(x) a_0(x) \frac{dy(x)}{dx}\right] + \mu(x) a_2(x) y(x) = 0. \tag{4.5.3}$$

4.5.2 Fourier Series

Let $f(x)$ be a piecewise continuous function defined over the interval, $0 \leq x \leq 2\pi$. Let $f(x)$ be periodic, i.e.,

$$f(x + 2\pi) = f(x). \tag{4.5.4}$$

Then the Fourier series for $f(x)$ is

$$f(x) = a_0 + \sum_{k=1}^{\infty} [a_k \cos(k \times z) + b_k \sin(kx)], \tag{4.5.5}$$

where

$$a_0 = \left(\frac{1}{2\pi}\right) \int_0^{2\pi} f(x)\, dx, \tag{4.5.6}$$

$$a_k = \left(\frac{1}{\pi}\right) \int_0^{2\pi} f(x) \cos(kx)\, dx, \tag{4.5.7}$$

$$b_k = \left(\frac{1}{\pi}\right) \int_0^{2\pi} f(x) \sin(kx)\, dx. \tag{4.5.8}$$

In the calculation of the Fourier series of a particular function $f(x)$, the following orthogonality relations may be of value

$$\int_0^{2\pi} \cos(kx) \sin(mx)\, dx = 0, \tag{4.5.9}$$

$$\int_0^{2\pi} \cos(kx) \cos(mx)\, dx = \int_0^{2\pi} \sin(kx) \sin(mx)\, dx$$
$$= \begin{cases} \pi, & \text{if } k = m, \\ 0, & \text{if } k \neq m. \end{cases} \tag{4.5.10}$$

Note that if we wish to change the interval in x, $[0, 2\pi]$, to an arbitrary interval, $[a, b]$, in $\tilde{x}$, then this can be done by the transformation

$$x = c_1 + c_2 \tilde{x}_1$$

where

$$c_1 = -\left(\frac{2\pi a}{b-a}\right), \quad c_2 = \frac{2\pi}{b-a}. \tag{4.5.11}$$

It should also be obvious that if $f(x)$ is even or odd, then the corresponding Fourier series will contain, respectively, only cosine or sine terms. For example, if

$$f(x) = x^2, \quad -\pi \leq x \leq \pi, \tag{4.5.12}$$

then its Fourier series is

$$f(x) = \frac{\pi^2}{3} + \sum_{k=1}^{\infty} (-1)^k \left(\frac{4}{k^2}\right) \cos(kx). \tag{4.5.13}$$

Similarly, for

$$f(x) = x, \quad -\pi < x < \pi, \tag{4.5.14}$$

the Fourier series is

$$f(x) = \sum_{k=1}^{\infty} (-1)^{k+1} \left(\frac{2}{k}\right) \sin(kx). \tag{4.5.15}$$

However, the function

$$f(x) = x(1-x), \quad 0 \leq x \leq 1, \tag{4.5.16}$$

is neither even nor odd over this interval. However, since it does have an odd extension to the full x-axis, it will have a sine Fourier expansion given by the expression:

$$f(x) = x(1-x)$$
$$= \left(\frac{2^{\frac{5}{2}}}{\pi^3}\right) \sum_{k=1}^{\infty} \frac{\sin(2k-1)\pi x}{(2k-1)^3}. \tag{4.5.17}$$

See Figure 4.7 for a sketch of this function.

Finally, while not generally useful for linear problems, the following *inner product formula* exists:

Let $f(x)$ and $g(x)$ be defined on the interval, $-\pi < x < \pi$, and have the representations

$$f(x) = a_0 + \sum_{k=1}^{\infty} [a_k \cos(kx) + b_k \sin(kx)], \tag{4.5.18}$$

$$g(x) = A_0 + \sum_{k=1}^{\infty} [A_k \cos(kx) + B_k \sin(kx)], \tag{4.5.19}$$

96 ■ Introduction to Qualitative Methods for Differential Equations

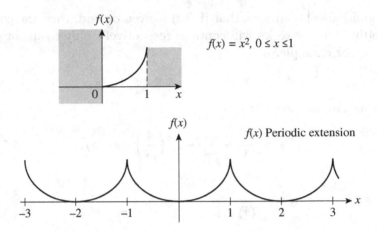

FIGURE 4.7 Sketches related to the periodic function $f(x) = x^2$.

then

$$\int_{-\pi}^{\pi} f(x)\, g(x)\, dx = 2\pi a_0 A_0 + \pi \sum_{k=1}^{\infty} [a_k A_k + b_k B_k]. \qquad (4.5.20)$$

4.5.3 Special Functions

There are a number of so-called 'special functions' that are solutions to the Sturm–Liouville problem. These functions have a wide applicability in the physical and engineering sciences. A partial listing of these items would include the following functions ($k = 0, 1, 2, ...$):

Hermite polynomials, $\{H_k(x)\}$, which appear in quantum systems involving the harmonic oscillator.

Legendre polynomials, $\{P_k(x)\}$, which occur in many problems in both classical and quantum physics.

Laguerre polynomials, $\{L_k(x)\}$, which are used to represent the radial part of hydrogen-type wavefunctions.

Chebyshev polynomials, $\{T_k(x), U_k(x)\}$ which are of great importance in both numerical analysis and approximation theory.

Jacobi polynomials, $\left\{P_k^{\alpha,\beta}(x)\right\}$ contain as special cases the Legendre, Chebyshev, and other orthogonal polynomials.

Bessel functions $\{J_k(x)$ *and* $Y_k(x)\}$ arise in many problems, classical and quantum, for which there is cylindrical or spherical symmetry.

A concise, but detailed summary of all the important features and properties of the Legendre, Chebyshev, Hermite and Laguerre functions are given in the book by Mickews (2017). See, in particular Sections 6.5.1 and the whole of Chapter 7. The full reference is

R. E. Mickens, *Mathematical Methods for the Natural and Engineering Sciences*, 2nd Edition (World Scientific, Singapore, 2017).

For each of these functions, the following items are given

- The differential equation
- Interval of definition and weight function
- Generating function
- Rodrique's formula (for generating the functions
- Orthogonality condition
- Recurrence relations
- List of first six functions in standard form
- Summary of special properties and value.

A discussion of Bessel functions is also included.

4.5.4 TISE: Sketches of Wavefunctions

Many important systems in quantum physics may be modeled as one-space dimension time-independent Schrödinger equations. These differential equations take the form

$$-\left(\frac{\hbar^2}{2m}\right)\frac{d^2\psi(x)}{dx^2} + V(x)\psi(x) = E\psi(x), \qquad (4.5.21)$$

with the requirement that $\psi(x)$ *is zero at the boundaries.* The $(\hbar, m)$ are atomic parameters, $V(x)$ is the potential energy function, assumed known, and E is to be determined. Thus, we have a Sturm–Liouville problem in which E and $\psi(x)$ play the roles of eigenvalue and eigenfunction.

Comments

(a) Our discussion is only for systems for which there are only bound states, i.e., the eigenvalues or eigen-energies take on only discrete values.

(b) There is also the requirement that if $\psi(x)$ is an eigenfunction, then it has the property

98 ■ Introduction to Qualitative Methods for Differential Equations

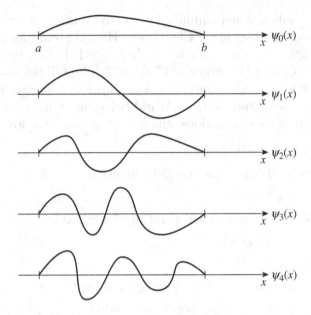

FIGURE 4.8 Generic behavior of the first five wavefunctions for the TISE defined in the interval, $a \le x \le b$.

$$\int_a^b \psi(x)^2 \, dx = 1. \tag{4.5.22}$$

(c) Note that for the situation under discussion, the differential equation for the TISE is defined over the interval

$$a \le x \le b, \tag{4.5.23}$$

with

$$\psi(a) = 0, \quad \psi(b) = 0. \tag{4.5.24}$$

(d) In the physical literature, $\psi(x)$ is called the wavefunction.

With these requirements in mind, let us now discuss/sketch the possible shapes of the wavefunctions for the following three cases for the location of the boundaries:

(1) $a \le x \le b$; both (a, b) bounded;
(2) $0 \le x < \infty$; a is finite and set to zero;

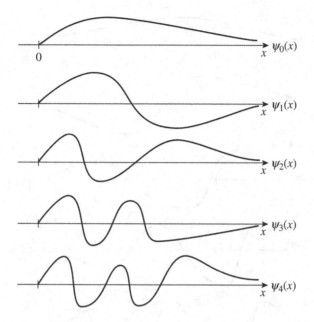

FIGURE 4.9 Generic behavior of the first five wavefunctions is for the TISE defined in the interval, $0 \leq x < \infty$. (For this case $a = 0$ and $b = \infty$.)

(3) $-\infty < x < \infty$; both boundaries are at infinity. In Figures 4.8 and 4.9, we will label the wavefunction for the smallest eigenvalue $\psi_0(x)$, in agreement with the usage in the physics community.

Figure 4.8 sketches these wavefunctions if the TISE is defined in the interval, $a \leq x \leq b$. A much used physical system modeled by such an equation is the particle-in-a-box.

A physical system having the wavefunctions given in Figure 4.9 is the interaction of two particles by means of a central force acting between them. For this case, x is the radial distance between them.

The wavefunctions sketched in Figure 4.10 are very similar to those that occur in the quantum modeling of a harmonic oscillator.

In Chapter 6, we will demonstrate how a knowledge of the general features of the wavefunctions can be used to construct analytic approximations to the wavefunctions of the TISE when exact solutions do not exist or are of such complex forms that we cannot usefully calculate with them.

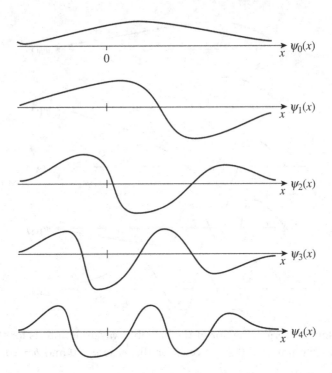

FIGURE 4.10 Generic behavior of the first five wavefunctiones for the TISE defined in the interval, $-\infty < x < +\infty$.

PROBLEMS

Section 4.1

1) Eliminate the first derivative term in the equation $y''(x) + 2 \in y'(x) + \omega_0^2 y(x) = 0$.

Explain the advantages of using the transformed equation.

2) Apply the Liouville–Green transformation to the equation $y''(x) + x^2 y = 0$. From this result, what general conclusions can be reached concerning the original differential equation?

Section 4.2

3) Apply and interpret the results obtained for the vibrating string to sounds produced by organ pipes.

Section 4.3

4) Show that if we have two sets of orthogonal eigenfunctions and their associated eigenvalues, i.e.,

$$[\lambda_k, \psi_k(x) \ : \ k = 0, 1, 2, ...] \text{ and } [\overline{\lambda}_m, \overline{\psi}_m(x) \ : \ m = 0, 1, 2, ...],$$

and if they satisfy the same boundary conditions, then each $\psi_k(x)$ can be expanded in terms of $\{\overline{\psi}_m(x)\}$ and vice versa.

COMMENTS AND REFERENCES

Sections 4.1–4.4: These sections are rewritten selections in from my book

(1) R. E. Mickens, *Mathematical Methods for the Natural and Engineering Sciences* (World Scientific, Singapore, 2017). See Chapter 4.

Two excellent texts on Slurm–Liouville problems that provide proofs of the quoted theorems are

(2) E. A. Coddington and N. Levinson, *Theory of Ordinary Differential Equations* (McGraw-Hill, New York, 1955).

(3) H. Sagan, *Boundary and Eigenvalue Problems in Mathematical Physics* (Wiley, New York, 1961).

CHAPTER 5

Partial Differential Equations

5.1 GENERAL COMMENTS

A partial differential equation is an expression that relates a function of several independent variables to its various partial derivatives. For the case of one-space coordinate, x, and time, t, some elementary examples are

$$u_{xt} = 0, \quad u_t + auu_x = bu_{xx}, \quad u_{tt} = c^2 u_{xx}, \qquad (5.1.1)$$

where $u = u(x,t), (a, b, c)$ are constants, and

$$u_x = \frac{\partial u(x,t)}{\partial x}, u_t = \frac{\partial u(x,t)}{\partial x}, u_{xt} = \frac{\partial^2 u(x,t)}{\partial x \partial t}, \quad \text{etc.} \qquad (5.1.2)$$

A partial differential equation (PDE) is linear if it is linear in the dependent variable and its various derivatives. Thus, in Equation (5.1.1), the first and third PDEs are linear, while the second is nonlinear because of the uu_x term.

The order of a PDE is the highest order of the derivatives that appear in the PDE. Therefore, in Equation (5.1.1), the first PDE is first order with respect to t and first order with respect to x; the second PDE is first order with respect to tt but second order with respect to x; and the third PDE is second order with respect to both x and t.

It should be noted that PDEs do not have general solutions in the sense that hold for ordinary differential equations; i.e., there exist functional relations such that all solutions are particular cases of these expressions. This fact makes the study and analysis of PDEs much harder to carry out than is the situation for ordinary differential equations.

One reason for this is that what appears in the solutions for ODEs as arbitrary constants shows up in PDEs as arbitrary functions of the independent variables. For example, consider the following linear second-order ODE

$$\frac{d^2 y(x)}{dx^2} = 0; \tag{5.1.3}$$

its general solution is

$$y(x) = A + Bx, \tag{5.1.4}$$

where (A, B) are arbitrary constants. However, the linear PDE

$$U_{xt}(x, t) = 0, \tag{5.1.5}$$

has a solution

$$u(x, t) = f(x) + g(t), \tag{5.1.6}$$

where $f(x)$ and $g(t)$ are arbitrary functions of x and t. It is only required that both functions have well-defined first derivatives.

Related to the discussion of the last several paragraphs is another important fact, i.e., often it is the particular or special solutions to PDEs that provide the solutions relevant to the study and investigation of many (if not most) physical phenomena, especially for nonlinear PDEs. Historically, two linear PDEs play critical roles in the modeling of physical systems, at least as good first approximations. These two equations are

$$u_t = D u_{xx}, \quad D > 0; \tag{5.1.7}$$

and

$$u_{tt} = c^2 u_{xx}, \quad |c| > 0;$$

and are called, respectively, the diffusion/heat and wave equations. Another important PDE that links these two equations is

$$\epsilon u_{tt} + u_t = D u_{xx}; \epsilon > 0, D > 0; \tag{5.1.8}$$

which in the research literature on heat conduction is called the Maxwell–Cattaneo equation; in mechanical engineering, it goes by the damped wave equation.

The purpose of this chapter is to examine several topics related to applying qualitative methods to the analysis of PDEs, with an emphasis on the diffusion/heat and wave equations and their generalizations. Section 5.2 demonstrates the use of symmetry arguments to derive these two PDEs, both in their linear and nonlinear manifestations. Next, in Section 5.3, we consider the method of separation-of-variables (MSOV) and show how it has been used to construct useful approximations to both linear and nonlinear PDEs. Finally, in section 5.4, we introduce the concept of traveling wave solutions (TWS) and calculate such solutions for several linear and nonlinear PDEs.

5.2 SYMMETRY-DERIVED PDES

Symmetry principles play important roles in the sciences, especially in physics. Many of the foundational PDEs may be derived just from the requirement that these equations must be invariant under particular symmetry transformations.

Further restrictions coming from an experimental knowledge of a system can place additional constraints on the sought-after mathematical structures. Also, limitations related to issues of what actually can be done experimentally can often be used to forbid certain mathematical terms or expressions.

The main goal of this section is to use symmetry arguments, along with measurability restrictions, to show that both the diffusion/heat and wave equations can be derived using these requirements. From these arguments, we also place limitations on possible nonlinear extensions of these two PDEs.

5.2.1 Heat Conduction PDE

One way to derive the heat conduction PDE is to assume that at the micro-level, the heat conduction is modeled by the following random walk equation

$$u(x, t+\tau) = \left(\frac{1}{2}\right) u(x+a, t) + \left(\frac{1}{2}\right) u(x-a, t), \qquad (5.2.1)$$

where the parameters, (a,τ), represent micro-level characteristic distance and time scales. This equation can be rewritten as

$$\frac{u(x,t+\tau)-u(x,t)}{\tau} = \left(\frac{a^2}{2\tau}\right)\left[\frac{u(x+a,t)-2u(x,t)+u(x-a,t)}{a^{(2)}}\right]. \tag{5.2.2}$$

If we now let

$$a \to 0, \tau \to 0, \text{ with } \frac{a^2}{2\tau} = D = \text{constant}, \tag{5.2.3}$$

then the following equation appears

$$\frac{\partial u(x,t)}{\partial t} = D\frac{\partial^2 u(x,t)}{\partial x^2}. \tag{5.2.4}$$

This is the standard diffusion/heat conduction PDE for one space dimension. The physical constant, D, has the physical units of (meter)^2/time.

To see what symmetry and other constraints give, let us make the following assumptions:

(i) We want a PDE that is linear and has constant coefficients.

(ii) The PDE must be invariant under $x \to -x$.

(iii) The PDE should not be invariant under $t \to -t$ (since it is to describe a dissipative system).

(iv) The PDE must satisfy a measurability condition. (This implies that time derivatives higher than the first order should not appear. This reflects the fact that at the current state of experimental methodology, these derivatives cannot be measured.)

A little thought, based on these four assumptions, leads to the following expression for the symmetry-derived PDE for the heat equation

$$u_t(x,t) = Du_{xx}(x,t) + (\tau D)u_{txx}(x,t). \tag{5.2.5}$$

COMMENTS

(1) The parameters (D,τ) in Equation (5.2.5) do not correspond to the same labelled parameters in Equations (5.2.1) and (5.2.4).

(2) However, in Equation (5.2.5), the two parameters (D, r), still have the physical units of (meter)/time, and time.

(3) Note that while Equation (5.2.5) is first order in the time derivative and second order in the space derivative, the second term on the right-hand side is of combined order three.

(4) Taking D and τ as just two physical parameters, the situation where $\tau = 0$ gives the standard heat PDE.

Nonlinear generalizations of the linear heat equation are easy to construct. For example, an expression consistent with the four assumption is

$$u_t = DF\left[(u_x)^2, u_{xx}\right] u_{xx} + (\tau D) G\left[(u_x)^2, u_{xx}\right] u_{txx}, \qquad (5.2.6)$$

where (F, G) are functions of the indicated derivatives. In particular, if we require an equation that is linear in u_{xx} and u_{txx}, the following is an example

$$u_t = D\left[1 + b_1 (u_x)^2\right] u_{xx} + (\tau D)\left[1 + b_2 (u_x)^2\right] u_{txx}^u, \qquad (5.2.7)$$

where now there are four physical parameters, (D, τ, b_1, b_2).

5.2.2 Wave PDE

The wave equation is

$$u_{tt}(x, t) = c^2 u_{xx}(x, t), \qquad (5.2.8)$$

where c is the characteristic velocity associated with the wave propagation, where $u(x, t)$ is the transverse displacement of a one-space dimension string at location x and time t. This can be demonstrated by defining $u_m(t)$ to be

$$u_m(t) = u(x_m, t), \ x_m = (\Delta x) m, \ m = \text{integers}, \qquad (5.2.9)$$

and discretizing u_{xx} as follows:

$$u_{xx} \to \frac{u_{m+1}(t) - 2u_m(t) + u_{m-1}(t)}{(\Delta x)^2}. \qquad (5.2.10)$$

Therefore, Equation (5.2.8) becomes

$$\frac{d^2 u_m}{dt^2} = \left[\frac{c^2}{(\Delta x)^2}\right](u_{m+1} - 2u_m + u_{m-1}). \qquad (5.2.11)$$

If all the 'particles' located at (x_m) have the same mass, M, then it follows that

$$M\frac{d^2 u_m}{dt^2} = k(u_{m+1} - u_m) - k(u_m - u_{m-1}), \qquad (5.2.12)$$

where

$$K = \frac{Mc^2}{(\Delta x)^2}. \qquad (5.2.13)$$

Thus, we can interpret Equation (5.2.12) as a chain of identical coupled simple harmonic oscillators having for each oscillator the effective mass, M, and a force constant, k, given in Equation (5.2.13).

Equation (5.2.11) can further be interpreted as a micro-level model for the continuum macroscopic level PDE represented by Equation (5.2.8).

Let us now derive a symmetry-based formulation of the wave equation. The two major requirements are the following:

(a) The PDE must be linear with constant coefficients.

(b) The PDE should be invariant under the transformations

$$x \to -x, \quad t \to -t. \qquad (5.2.14)$$

The lowest-order PDE satisfying these conditions is

$$U_{tt} = au_{xx}, \quad a > 0, \qquad (5.2.15)$$

a is a positive parameter having the physical units of (meters/time)^2, i.e., the units of speed.

If we consider nonlinear versions of the wave equation that contain only derivative terms, then a generalization of the linear wave equation takes the form $(a = c^2)$

$$u_{tt} = c^2 F\left[(u_t)^2, (u_x)^2\right] u_{xx}, \qquad (5.2.16)$$

where
$$F(0,0) = 1. \tag{5.2.17}$$

If $F(v, w)$ is assumed to have a Taylor expansion at $(v, w) = (0, 0)$, then
$$F(v, w) = 1 + f_1 v + f_2 w + O(v^2) + O(w^2), \tag{5.2.18}$$

where f_1 and f_2 are constants, and this leads to the nonlinear wave equation
$$u_{tt} = c^2 \left[1 + f_1 (u_t)^2 + f_2 (u_x)^2 \right] u_{xx}. \tag{5.2.19}$$

5.2.3 Discussion

The four broadly used, linear, constant coefficient PDEs appearing in the natural and engineering sciences are

- Diffusion/heat equation operator
$$u_t = D \Delta^2 u \tag{5.2.20}$$

- Wave equation
$$u_{tt} = c^{(2)} \Delta^2 u \tag{5.2.21}$$

- Laplace's equation
$$\Delta^2 u = 0 \tag{5.2.22}$$

- Poisson's equation
$$\Delta^2 u = \rho \tag{5.2.23}$$

where Δ^2 is the Laplace operator, which can be expressed in Cartesian coordinates as
$$\Delta^2 \equiv \frac{\partial^2}{\partial x^2} + \frac{\partial^2}{\partial y^2} + \frac{\partial^2}{\partial z^2}. \tag{5.2.24}$$

In the above equations, D and c^2 are taken to be constants, and ρ is a function of the space coordinates. However, in a fundamental sense, only the wave and diffusion/heat equations are essential. This is because if we write out the most complete and general linear wave PDE, i.e.,

$$\epsilon u_{tt} + a u_t = b\Delta^2 u + \vec{v}_0 \Delta u + \rho, \qquad (5.2.25)$$

where $\vec{v}_0$ is a constant vector, and (ϵ, a, b) are constants, then all of the other three PDEs are just special cases.

Note that the symmetry arguments, given above, suggest that the elementary forms of the wave and diffusion/heat PDEs are fundamental and are expected to show themselves in all branches of the natural and engineering sciences. The main reason for our making this statement is that the considered symmetries are direct consequences of an analysis of both experimental data and the requirements needed to actually carry out a valid scientific experiment, i.e., in more detail:

(i) When we initiate an experiment should play no role in determining the obtained results. (The associated symmetry transformation is 'invariance' under $t \to t_0 + t$, where t_0 is arbitrary.)

(ii) The location of the experiment is irrelevant. (The symmetry transformation is 'invariance' under $x \to x_0 + x$, where x_0 is arbitrary.)

(iii) The choices made for the orientation and positive directions of the coordinates should not matter.

(iv) Constructed mathematical models and theories should only include parameters and functions that can be actually measured.

COMMENTS

(1) The parameters, (D, τ) in Equation (5.2.5) do not correspond to the same labeled parameters in Equations (5.2.1) and (5.2.4).

(2) However, in Equation (5.2.5), the two parameters, (D, r), still have the physical units of (meter)/time and time.

(3) Note that while Equation (5.2.5) is first order in the time derivative and second order in the space derivative, the second term on the right-hand side is of combined order 3.

(4) Taking D and τ as just two physical parameters, the situation where $\tau = 0$ gives the standard heat PDE.

Nonlinear generalizations of the linear heat equation are easy to construct. For example, an expression consistent with the four assumptions is

$$u_t = DF\left[(u_x)^2, u_{xx}\right] u_{xx}$$
$$+ (\tau D)\, G\left[(u_x)^2, u_{xx}\right] u_{txx}, \qquad (5.2.6)$$

where (F, G) are functions of the indicated derivatives. In particular, if we require an equation that is linear in u_{xx} and u_{txx}, the following is an example

$$u_t = D\left[1 + b_1(u_x)^2\right] u_{xx} + (\tau D)\left[1 + b_2(u_x)^2\right]_{txx}^u, \qquad (5.2.7)$$

where now there are four physical parameters, (D, τ, b_1, b_2).

5.3 METHOD OF SEPARATION OF VARIABLES

5.3.1 Introduction

Partial differential equations do not have general solutions in the sense of ordinary differential equations. PDEs may have various classes of solutions, which cannot be reduced to each other. This result is a consequence of the fact that arbitrary functions may appear in the solutions of PDEs. For example, the linear PDE

$$u_{xt}(x, t) = 0, \qquad (5.3.1)$$

has the solution

$$u(x, t) = f(x) + g(t), \qquad (5.3.2)$$

where $f(x)$ and $g(t)$ are arbitrary except for the requirement that they each have a first derivative. The usual case is that for any given PDE, whether linear or nonlinear, only particular or special solutions can be found.

The main purpose of this section is to introduce a methodology for determining special solutions to PDEs. But, this technique, named the method of separation-of-variables (SOV), only applies to a limited type of PDEs, and therefore it is not of general applicability. This procedure can also be applied to some ODEs.

The method of SOV is based on the realization that if we can find any solution that solves the PDE and, in addition, satisfies the initial and/or boundary conditions for the problem under study, then this is the required solution.

In the next subsection, we define the method of SOV and discuss some of its advantages and disadvantages. This is followed by subsections where we apply this technique, respectively, to ODEs and PDEs.

5.3.2 Definition of the Method of SOV

Consider an ODE that has the structure

$$\frac{dy}{dx} = f(x)\,g(y). \tag{5.3.3}$$

Note that the derivative is equal to a product of two functions, one depending on the independent variable and the other depending only on the dependent variable. We call this ODE separable and its general solution is

$$\int \frac{dy}{g(y)} = \int f(x)\,dx + C. \tag{5.3.4}$$

An elementary example is

$$\frac{dy}{dx} = -\frac{x}{y}, \tag{5.3.5}$$

where

$$f(x) = -x, \quad g(y) = \frac{1}{y}. \tag{5.3.6}$$

Therefore,

$$\int y\,dy = -\int x\,dx, \tag{5.3.7}$$

and we obtain

$$y^2 + x^2 = r^2, \quad C = r^2, \tag{5.3.8}$$

consequently, there are two functions that are solutions

$$y_+(x) = \sqrt{r^2 - x^2}, \quad y_-(x) = -\sqrt{r^2 - x^2}. \tag{5.3.9}$$

In general, there may exist other solutions to Equation (5.3.3) in addition to the one given by Equation (5.3.4). The other solutions are constant functions and are called fixed-point or equilibrium solutions. They correspond to the (real) zeros of $g(y)$, i.e.,

$$y_i(x) = \bar{y}_i \; : \; g(\bar{y}_i) = 0, i = (1, 2, ..., I). \tag{5.3.10}$$

Thus, for

$$\frac{dy}{dx} = xy(1-y), \tag{5.3.11}$$

we have

$$f(x) = x, \quad g(y) = y(1-y), \tag{5.3.12}$$

with

$$g(\bar{y}) = 0 \; : \; \bar{y}_1 = 0, \bar{y}_2 = 1. \tag{5.3.13}$$

Consequently, the solution to Equation (5.3.11) consists of three functions

$$\int \frac{dy}{y(1-y)} = \frac{x^2}{2} + C, \quad y_2(x) = 0, \quad Y_3(x) = 1. \tag{5.3.14}$$

For PDEs the situation is different. For this case, we assume that the solution is written as a product of functions, each depending on one independent variable. So if $u = u(x, t)$, the separation of the variable solution is

$$u(x, t) = F(x) G(t). \tag{5.3.15}$$

If this ansatz 'works', then two separate ODEs will be obtained for the functions $F(x)$ and $G(t)$. In essence, the original PDE has been reduced to a set of ODEs. The ODEs are related to each other by a so-called constant of separation. For the situation where the PDE is linear, we can add solutions for different values of the separation constant to obtain a general solution.

Since PDE problems generally come with certain initial conditions and boundary restrictions, we should make sure that our functions satisfy the boundary conditions, otherwise, the method of SOV fails. It must be kept in mind that the method of SOV does not always work, but when it does, it provides very satisfactory results.

We now illustrate the value of this method by considering several ODEs and PDEs.

5.3.3 Examples

Example 5.1 *The first-order, nonauton ODE*

$$\frac{dy}{dx} = 2x(1-y)^2, \qquad (5.3.16)$$

is separable and has an equilibrium solution

$$y_1(x) = 1. \qquad (5.3.17)$$

The other, one parameter, solution can be obtained from the expression

$$\int \frac{dy}{(1-y)^2} = \int 2x\,dx + C, \qquad (5.3.18)$$

where C is an arbitrary constant. Integrating this gives

$$\frac{1}{1-y} = x^2 + C, \qquad (5.3.19)$$

which when solved for y gives

$$y_2(x) = 1 - \frac{1}{x^2 + C} \qquad (5.3.20)$$

If we require $y(0) = y_0$, the C can be calculated; it is

$$C = \frac{1}{1-y_0}, \qquad (5.3.21)$$

and this gives for Equation (5.3.18) the result

$$y_2(x) = \frac{(1-y_0)x^2 + y_0}{(1-y_0)x^2 + 1}. \qquad (5.3.22)$$

Observe that for $y_0 > 1$, $y_2(x)$ has singularities at x_s, where

$$x_s = (\pm)\frac{1}{\sqrt{y_0 - 1}} \qquad (5.3.23)$$

Also, an inspection of $y_2(x)$ shows that there is no finite value of C for which $y_1(x) = 1$ is a solution. The result follows from the facts that the integrand for the y integral is not defined for $y = 1$; see the right-hand side of Equation (5.3.17).

Example 5.2 *Let us solve and analyze the solutions for*

$$\frac{dy}{dx} = \left(\frac{1}{2}\right) x(1 - y^2). \tag{5.3.24}$$

First, there are two equilibrium or constant solutions; they are

$$y_1(x) = +1, \, y_2(x) = -1. \tag{5.3.25}$$

Since this ODE is separable, we can integrate it in the form

$$\int \frac{2\,dy}{(1 - y^2)} = \int dx + C. \tag{5.3.26}$$

If we use

$$\frac{2}{y^2 - 1} = \left(\frac{1}{y - 1}\right) - \left(\frac{1}{y + 1}\right), \tag{5.3.27}$$

and do the integrations in Equation 5.3, the following result is obtained

$$\mathrm{Ln}\left|\frac{y-1}{y+1}\right| = -\left(\frac{x^2}{z}\right) + C, \tag{5.3.28}$$

or solving for $y(x)$ or $y_3(x)$,

$$y_3(x) = \frac{1 + ce^{-\frac{x^2}{z}}}{1 - ce^{-\frac{x^2}{z}}}, \quad y^2 \neq 1, \tag{5.3.29}$$

where the two C's in Equations (5.3.28) and (5.3.29) are not the same, but are related to each other. The solution to Equation (5.3.24) is composed of the three functions given in Equations (5.3.25) and (5.3.29).

Another way to analyze the behavior of the solutions to this differential equation is to examine the integral curves for Equation (5.3.24) in the x–y plane. Observe that the ODE is invariant under the transformation

$$x \to -x, \quad y \to y. \tag{5.3.30}$$

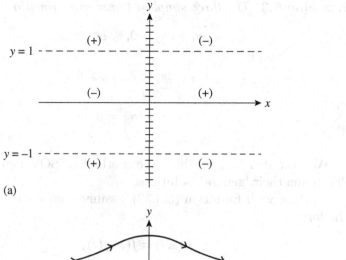
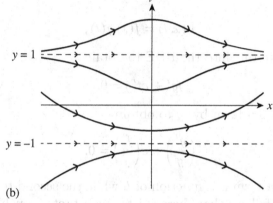

FIGURE 5.1 (a) The domains of constant y' signs for $y' = x(1-y^2)/2$. (b) Typical solution curves in the $x-y$ plane.

This means that in the $x-y$ plane, the solutions or integral curves are symmetric with respect to the y-axis. Also, further inspection of the ODE shows that the derivative is zero along the curves

$$y' = 0 \; : \; \begin{cases} x = 0 \text{ or the } y-\text{axis}; \\ y_0(x) = (\pm)\,1. \end{cases} \qquad (5.3.31)$$

Figure 5.1 provides a representation of the $x-y$ plane and the six domains where the sign of the derivative, $y' = dy(x)/dx$, is either plus or minus. It also sketches several solution or integral curves, i.e., $y(x)$ vs x.

Example 5.3 *The three simplest linear wave equations are*

$$u_t + cu_x = 0, \quad c > 0; \tag{5.3.32}$$

$$u_t - cu_x = 0, \quad c > 0; \tag{5.3.33}$$

$$u_{tt} - c^2 u_{xx} = 0. \tag{5.3.34}$$

We now demonstrate how the method of SOV can be used to determine their 'general' solutions.

Starting with Equation (5.3.32), assume that it has a solution of the form

$$u(x,t) = f(x)g(t), \tag{5.3.35}$$

and substitute this into the PDE to obtain

$$fg' + cf'g = 0. \tag{5.3.36}$$

Next divide each term by fg to obtain

$$\left(\frac{g'}{g}\right) + c\left(\frac{f'}{f}\right) = 0. \tag{5.3.37}$$

Since the first term is a function of t, while the second term depends only on x, it follows that there exists a constant, k, such that

$$\frac{g'}{cg} = -k, \quad \frac{f'}{f} = +k. \tag{5.3.38}$$

These first-order, linear ODEs have solutions

$$g(t) = A(k)e^{-kct}, \quad f(x) = B(k)e^{kx}, \tag{5.3.39}$$

where, for the moment, $A(k)$ and $B(k)$ are arbitrary functions of k. Therefore,

$$u(x,t,k) = F(k)e^{k(x-ct)}, \quad F(k) = A(k)B(k), \tag{5.3.40}$$

for some k is a solution of Equation (5.3.32).

If we wish to have bounded solutions, then we can select k to be purely imaginary, i.e., write

$$k \to ik, k(\text{on the right-side real}). \tag{5.3.41}$$

With this change, Equation (5.3.40) becomes

$$u(x, t, k) = F(k) e^{ik(x-ct)}, \tag{5.3.42}$$

where the 'i' is not indicated in the functions u and F. Summing over discrete k and integrating over real, continuous k, we finally obtain

$$\begin{aligned} u(x, t) &= \int_{-\infty}^{\infty} F(k) e^{ik(x-ct)} dk \\ &= H(x - ct), \end{aligned} \tag{5.3.43}$$

i.e., $u(x, t)$ is just a function of $(x - ct)$. It is assumed that $F(k)$ has properties such that the sum/integration exists.

In summary, the PDE, $u_t + cu_x = 0$, has as a solution an arbitrary function of $Z = x - ct$, where this function possesses a first derivative. We can check this by observing that

$$\begin{cases} \frac{\partial}{\partial t} H(x - ct) = -cH', & \frac{\partial}{\partial x} H(x - ct) = H', \\ H'(z) = \frac{dH(z)}{dz}. \end{cases} \tag{5.3.44}$$

This technique can be applied to Equation (5.3.34) to show that the solution to it is

$$u_t - cu_x = 0 \Rightarrow u(x, t) = M(x + ct), \tag{5.3.45}$$

where $M(w)$ is an arbitrary function of $w = x + ct$, where dM/dw exists.

Also, since

$$u_{tt} - c^2 u_{xx} = (\partial_t - c\partial_x)(\partial_t + c\partial_x) u = 0, \tag{5.3.46}$$

it follows that the solution is

$$U(x, t) = H(x - ct) + M(x + ct). \tag{5.3.47}$$

Example 5.4 *The following nonlinear PDE is a type of nonlinear diffusion equation*

$$u_t = uu_{xx}, \quad u = u(x, t). \tag{5.3.48}$$

Let us find an SOV solution of the form

$$u(x, t) = f(x) g(t). \qquad (5.3.49)$$

Substituting this expression into the PDE gives

$$fg' = (fg)(f''g) \qquad (5.3.50)$$

or

$$\frac{g'}{g^2} = f'' = -\lambda = \text{separation constant}. \qquad (5.3.51)$$

Therefore, $g(t)$ and $f(x)$ satisfy the ODEs

$$\frac{dg(t)}{dt} = -\lambda g^2(t), \quad \frac{d^2 f}{dx^2} = -\lambda, \qquad (5.3.52)$$

with the respective solutions

$$g(t) = \frac{1}{A + \lambda t}, \quad f(x) = -\left(\frac{\lambda}{2}\right) x^2 + B_1 x + B_2. \qquad (5.3.53)$$

Therefore, the SOV solution for the PDE, $u_t = u u_{xx}$, is

$$u_{\text{SOV}}(x, t) = \frac{-\left(\frac{\lambda}{2}\right) x^2 + B_1 x + B_2}{A + \lambda t}, \qquad (5.3.54)$$

where (λ, A, B_1, B_2) are arbitrary constants.

Note that

$$u(x, t) = c_1 + c_2 x, \qquad (5.3.55)$$

is the solution to the equilibrium, steady state of a system modeled by Equation (5.3.48), i.e., $u_{xx} = 0$.

Example 5.5 *The Burgers' equation is*

$$u_t + u u_x = D u_{xx}. \qquad (5.3.56)$$

However, our interest is for the case where $D = 0$. This is the diffusion-free Burgers' equation, i.e.,

$$u_t + u u_x = 0. \qquad (5.3.57)$$

Assuming $u(x,t) = f(x)g(t)$ and substituting this into Equation (5.3.57) give

$$fg' + (fg)(f'g) = 0 \tag{5.3.58}$$

which if now is divided by fg yields the expression

$$\frac{g'}{g^2} + f' = 0, \tag{5.3.59}$$

or

$$g' = \lambda f^2, \quad f' = -\lambda. \tag{5.3.60}$$

The solutions to these ODEs are

$$g(t) = \frac{1}{A_1 - \lambda t}, f(x) = A_2 - \lambda x, \tag{5.3.61}$$

where (λ, A_1, A_2) are arbitrary constants. Therefore, the SOV solution for the diffusionless Burgers' equation is

$$u_{\text{SOV}}(x,t) = \frac{A_2 - \lambda x}{A_1 - \lambda t}. \tag{5.3.62}$$

Example 5.6 *Simple heat conduction in one dimension is usually studied by application of the so-called heat equation*

$$\frac{\partial u(x,t)}{\partial t} = D\frac{\partial^2 u(x,t)}{\partial x^2}, \quad D > 0. \tag{5.3.63}$$

A generalization of this equation is

$$\frac{\partial u}{\partial t} + k(t)u = f(x)\frac{\partial^2 u}{\partial x^2} + g(x)\frac{\partial u}{\partial x} + h(x)u, \tag{5.3.64}$$

where our major requirement is that $f(x) > 0$ for all relevant values of x, i.e.,

$$a \le x \le b. \tag{5.3.65}$$

Also, we impose the following boundary and initial conditions

$$u(a,t) = 0, \quad u(b,t) = 0, \quad u(x,0) = S(x) \tag{5.3.66}$$

Let us assume that an SOV solution exists, i.e.,

$$u(x, t) = X(x) T(t). \qquad (5.3.67)$$

Also, define the two operators, $\hat{L}(x)$ and $\hat{M}(t)$, as

$$\begin{cases} \hat{L} = \hat{L}(x) \equiv f(x)\dfrac{d^2}{dx^2} + g(x)\dfrac{d}{dx} + h(x), \\ \hat{M} = \hat{M}(t) \equiv \dfrac{d}{dt} + K(t). \end{cases} \qquad (5.3.68)$$

If we substitute $u = XT$ into our PDE and rearrange its various terms, then we obtain the result

$$\frac{\hat{L}X}{X} = \frac{\hat{M}T}{T}. \qquad (5.3.69)$$

Since the left-hand and right-hand sides are, respectively, functions only of x and t, then they must be equal to the same constant, which we label, λ. Therefore, Equation (5.3.69) becomes the two equations

$$\begin{cases} \hat{L}(x) X(x) = \lambda X(x); X(a) = 0, X(b) = 0, \\ \hat{M}(t) T(t) = \lambda T(t), T(0) \text{ arbitrary.} \end{cases}$$

Note that the first of these equations can be treated as a Sturm–Liouville problem, and this will give us a set of orthonormal eigenfunctions and eigenvalues, i.e.,

$$\hat{L}X = \lambda X \rightarrow [(X_n(x), \lambda_n) \; : \; n = 1, 2, 3, ...], \qquad (5.3.70)$$

with

$$\int_a^b X_n(x) X_m(x) w(x) \, dx = \delta_{nm}, \qquad (5.3.71)$$

and $w(x)$, the weight function, is given by

$$w(x) = \left[\frac{1}{f(x)}\right] \exp\left[\int \frac{g(x)}{f(x)} dx\right]. \qquad (5.3.72)$$

Once the eigenvalues, $(\lambda_n \; : \; n = 1, 2, 3, ...)$ are determined, then the $T(t)$ functions can be calculated from the following linear, first-order ODE

$$\frac{dT_n(t)}{dt} = [\lambda_n - k(t)] T_n(t). \qquad (5.3.73)$$

The solution to this equation can be written as

$$T_n(t) = \exp\left[\lambda_n t - \int_0^t k(z)\,dz\right], \qquad (5.3.74)$$

where we have used $T_n(0) = 1$.

Finally, with all of these results, the SOV solution to the initial/boundary value problem, defined by Equations (5.3.62)–(5.3.64), is

$$u(x,t) = \sum_{n=1}^{\infty} a_n X_n(x)\, T_n(t), \qquad (5.3.75)$$

where

$$a_n = \int_a^b S(x)\, X_n(x)\, w(x)\, dx. \qquad (5.3.76)$$

Note that the a_n are the expansion coefficients for the initial function, $S(x)$, in terms of the eigenfunctions, i.e.,

$$u(x,0) = S(x) = \sum_{n=1}^{\infty} a_n X_n(x). \qquad (5.3.77)$$

Example 5.7 *Many fundamental physical systems can be modeled within the context of three space dimensions. An exceptionally important example is the quantum model of the hydrogen atom. For this and also many classical systems, the modeling PDEs can be separated into three ODEs by application of the method of SOV. In general, these ODEs are the differential equations for special functions such as the*

- *Trigonometric and hyperbolic functions*
- *Bessel functions*
- *Jacobi elliptic functions*
- *Hermite polynomials*
- *Laguerre polynomials*
- *Chebyshev polynomials*
- *Legendre functions*
- *Lambert W-function*

The method of SOV plays a prominent role in the quantum model and its interpretation of the structure of the hydrogen atom. The details of this mathematical model and the derived physical properties are given in many textbooks. A concise summary is provided at https://web.mst.edu/~sparlin/phys107/lecture/chap06.pdf.

The Schrödinger equation for the hydrogen atom in spherical coordinates, (r, θ, φ), is

$$-\left(\frac{\hbar^2}{2m}\right)\Delta^2\psi + U(r)\psi = E\psi, \psi = \psi(r,\theta,\varphi), \quad (5.3.78)$$

where

$m = \dfrac{m_e m_p}{m_e + m_p}$; m_e = electron mass, m_p = proton mass;

$\hbar$ = reducedPlanck constant;

$U(r) = -\left(\dfrac{ke^2}{r}\right)$, potential energy of interaction between the charge (-e) of the electron and the charge (+e) of the proton;

r = distance between the electron and proton;

E = energy eigenvalues;

ψ = wave function eigen functions.

Written in full spherical coordinates, Equation (5.3.78) is the expression

$$\left(\frac{1}{r^2}\right)\frac{\partial}{\partial r}\left(r^2\frac{\partial \psi}{\partial r}\right) + \left(\frac{1}{r^2 \sin\theta}\right)\frac{\partial}{\partial \theta}\left(\sin\theta\frac{\partial \psi}{\partial \theta}\right)$$
$$+ \left(\frac{1}{r^2 \sin\theta}\right)\frac{\partial^2 \psi}{\partial \varphi^2} + \left(\frac{2m}{\hbar^2}\right)(E-U) = 0. \quad (5.3.79)$$

If we now use an SOV ansatz for ψ, i.e.,

$$\psi(r,\theta,\varphi) = R(r)\Theta(\theta)\Phi(\varphi), \quad (5.3.80)$$

and then use separation constants to isolate each independent variable, the following three ODEs arise

$$\frac{d^2\Phi}{d\varphi^2} + m_l^2 \Phi = 0, \quad (5.3.81)$$

$$\left(\frac{1}{\sin\theta}\right)\frac{d}{d\theta}\left(\sin\theta\frac{d\Theta}{d\theta}\right) + \left[l(l+1) - \frac{m_l^2}{(\sin\theta)^2}\right]\Theta = 0, \quad (5.3.82)$$

$$\left(\frac{1}{r^2}\right)\frac{d}{dr}\left(r^2\frac{dR}{dr}\right) + \left[\left(\frac{2m}{\hbar^2}\right)(E-U) - \frac{l(l+1)}{r^2}\right]R = 0. \qquad (5.3.83)$$

Note that these three ODEs are associated, respectively, with the defining differential equations of the trigonometric, Legendre, and Laguerre functions.

5.4 TRAVELING WAVES

Definition 1 *A PDE of the form*

$$u_t = H(u, u_x, u_{xx}, \ldots), \quad u = u(x,t), \qquad (5.4.1)$$

with a maximum order in the x-derivative is called an evolution PDE.

Definition 2 *A traveling wave solution of a PDE is any solution that can be represented as*

$$u(x,t) = f(x - ct), \qquad (5.4.2)$$

with c constant.

The traveling wave solutions are generally special solutions to a PDE. However, they play an important role in the natural and engineering sciences because they appear in a wide variety of phenomena. Examples extend from pulses in fiber optics to electrical propagation along nerves. Again, these are illustrations of cases where special solutions are the required solutions, rather than more general mathematical representations.

Since traveling wave solutions play such important roles in the sciences, a vast research literature exists on techniques to calculate them for particular classes of PDEs. The derived methods allow the determination (sometimes) of exact and/or approximate traveling wave solutions. A short listing of some of the recent works on this topic includes the following items:

1. W. Malfliet and W. Hereman, The tanh method: I. Exact solutions of nonlinear evolution and wave equations, *Physica Scripta*, Vol. 54, No. 6 (1996), 563–568.

2. D. Bazeia, A. Das, L. Losano, and A. Silva, A simple and direct method for generating travelling wave solutions for nonlinear equations, *Annuals of Physics*, Vol. 323, No. 5 (2008), 1150–1167.

3. D. Bazeia, A. Das, L. Losano, and M. J. Santos, Traveling wave solutions of nonlinear partial differential equations, *Applied Mathematics Letters*, Vol. 23, No. 6 (2010), 681–686.
4. G. W. Griffiths and W. E. Schiesser, *Traveling Wave Analysis of Partial Differential Equations* (Elsevier, Amsterdam, 2012).
5. S. W. Cho, H. J. Hwang, and H. Son, Traveling wave solutions of partial differential equations via neural networks, *Journal of Scientific Computing*, Vol. 89 (2021), Article Number 21.

We give in the remainder of this section four PDEs for which explicit formulas can be calculated for their traveling wave solutions. However, we note that the existence and calculation of such solutions for a particular PDE is often a long and complex process if it can be done at all.

5.4.1 Burgers' Equation

The standard form of this PDE is

$$u_t + uu_x = Du_{xx}, \quad D > 0. \tag{5.4.3}$$

Substituting

$$u(x,t) = f(x - ct) \tag{5.4.4}$$

into the PDE gives

$$-cf' + ff' = Df'', \tag{5.4.5}$$

where

$$z = x - ct, \quad f'(z) = \frac{df(z)}{dz}. \tag{5.4.6}$$

If Equation (5.4.5) is integrated once and the terms are rearranged, then we obtain the expression

$$\frac{df}{dz} = \left(\frac{1}{2D}\right)(f^2 - 2cf + 2A), \tag{5.4.7}$$

where A is (for now) an arbitrary integrations constant.

Considering the whole z-axis, i.e., $-\infty < z < +\infty$, let us require the following conditions to hold,

$$\lim_{z \to -\infty} f(z) = u_1, \quad \lim_{z \to +\infty} f(z) = u_2, \, u_1 > u_2 > 0, \tag{5.4.8}$$

where (u_1, u_2) are the constants. With this information, we can write

$$\frac{df}{dz} = \left(\frac{1}{2D}\right)(f - u_1)(f - u_2) \tag{5.4.9}$$

$$= \left(\frac{1}{2D}\right)[f^2 - (u_1 + u_2)f + u_1 u_2]. \tag{5.4.10}$$

Comparing Equations (5.4.7) and (5.4.10) allows for the determination of the speed, c, and the constant, A; they are

$$A = \frac{u_1 u_2}{2}, \quad c = \frac{u_1 + u_2}{2}. \tag{5.4.11}$$

Using the result of Equation (5.4.10), the general behavior of $f(z)$ is sketched in Figure 5.2.

Equation (5.4.10) is a separable ODE and can be solved for its solution, which is

$$f(z) = u_2 + \frac{(u_1 - u_2)}{1 + \exp\left[\left(\frac{u_1 - u_2}{2D}\right)z\right]}. \tag{5.4.12}$$

Observe that taking the limit, $D \to 0$, and being careful, we find that Equation (5.4.12) reduces to a *shock wave*, i.e.,

$$f(z) = u_1 \theta(z) + u_2 \theta(-z), \tag{5.4.13}$$

where

$$\theta(z) = \begin{cases} 1, & z < 0; \\ 0, & z > 0, \end{cases} \tag{5.4.14}$$

A sketch of the shock wave solution is also given in Figure 5.2.

5.4.2 Korteweg de Vries Equation

This PDE is

$$u_t + u u_x + u_{xxx} = 0. \tag{5.4.15}$$

We wish to seek traveling wave solutions

$$u(x, t) = f(z), \quad z = x - ct, \tag{5.4.16}$$

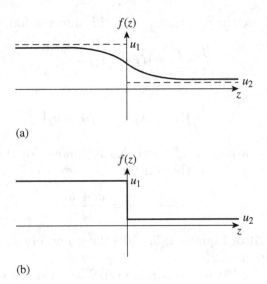

(a)

(b)

FIGURE 5.2 Sketches of Equations (5.4.11): (a) $D > 0$; (b) $D = 0$.

where $f(z)$ has the properties

$$\lim_{|z|\to\infty} f^{(k)}(z) = 0, \, k = (0, 1, 2, 3). \tag{5.4.17}$$

Substitution of Equation (5.4.16) into Equation (5.4.15) gives

$$-cf' + ff' + f''' = 0. \tag{5.4.18}$$

Integrating once, we obtain

$$-cf + \left(\frac{1}{2}\right)f^2 + f'' = 0. \tag{5.4.19}$$

If this equation is multiplyed by f', i.e.,

$$-c(ff') + \left(\frac{1}{2}\right)(f^2 f') + f'f'' = 0, \tag{5.4.20}$$

then it can be integrated to give

$$-\left(\frac{c}{2}\right)f^2 + \left(\frac{1}{6}\right)f^3 + \left(\frac{1}{2}\right)(f')^2 = 0. \tag{5.4.21}$$

Solving for f' with the positive root selected gives

$$\frac{\sqrt{3}}{\sqrt{3c-f}} \cdot f' = 1. \tag{5.4.22}$$

From the appearance of the square-root term, it follows that a real solution will only exist if the following restriction holds

$$0 \le f(z) \le 3c. \qquad (5.4.23)$$

If the change of variables

$$g^2(z) = 3c - f(z) \qquad (5.4.24)$$

is made, then we obtain from Equation (5.4.22) the expression

$$\left(\frac{2\sqrt{3}}{3c - g^2}\right) g' = 1, \qquad (5.4.25)$$

and this ODE can be solved to obtain

$$g(z) = -\left(\sqrt{3}\,c\right) \tanh\left[\left(\frac{1}{2}\right)\sqrt{c}\,z\right], \qquad (5.4.26)$$

where the integration constant has been set to zero. Using Equation (5.4.24), $f(z)$ is

$$f(z) = (3c)\ \text{sech}^2\left[\left(\frac{1}{2}\right)\sqrt{c}\,z\right], \qquad (5.4.27)$$

and the traveling wave solution is

$$u(x,t) = f(x - ct) = (3c)\ \text{sech}^2\left[\left(\frac{\sqrt{c}}{2}\right)(x - ct)\right]. \qquad (5.4.28)$$

The sech(y) function is

$$\text{sech}(y) = \frac{2}{e^y + e^{-y}}, \qquad (5.4.29)$$

therefore, the traveling wave solution is a 'pulse' moving in the positive x-direction, if $c > 0$, with a speed c and a maximum height of $3c$. An important feature of this traveling wave is that higher pulses travel faster than smaller amplitude pulses.

It should be indicated that the *KdV* equation has many other types of solutions other than the single pulse that was discussed above. The book *Nonlinear Partial Differential Equations for Scientists and Engineering* by L. Debnath provides an excellent introduction to these issues. We now show that rational solutions exist for this PDE.

Assume that $u(x,t) = f(x)g(t)$ and substitute this into the *KdV* equation to obtain

$$fg' + (fg)(f'g) + f'''g = 0. \tag{5.4.30}$$

Dividing by fg gives

$$\frac{g'}{g} + f'g + \frac{f'''}{f} = 0 \tag{5.4.31}$$

Let

$$f''(x) = 0 \rightarrow f(x) = A + Bx, f'(x) = B, \text{ and } f'''(x) = 0. \tag{5.4.32}$$

Therefore,

$$\frac{g'}{g} + Bg = 0 \rightarrow g(t) = \frac{1}{Bt + C}, \tag{5.4.33}$$

and

$$u(x,t) = f(x)g(t) = \frac{A + Bc}{Bt + C}. \tag{5.4.34}$$

5.4.3 Fisher's Equation

The Fisher PDE

$$u_t = Du_{xx} + \lambda u(1-u), \quad 0 \le u \le 1, \tag{5.4.35}$$

may be the most famous of the so-called reaction–diffusion PDEs. The three terms in this equation have the following interpretations

$$\begin{pmatrix} \text{Evolution} \\ \text{of the} \\ \text{system} \end{pmatrix} = (\text{Diffusion}) + \begin{pmatrix} \text{Reaction} \\ \text{term} \end{pmatrix}.$$

For the remainder of this subsection, we will take $D = 1$ and $\lambda = 1$, thus giving the normalized expression

$$u_t = u_{xx} + u(1-u). \tag{5.4.36}$$

Also, an inspection of this equation shows that there are two equilibrium or constant solutions

$$\bar{u}^{(1)}(x,t) = 0, \quad \bar{u}^{(2)}(x,t) = 1, \tag{5.4.37}$$

and they are, respectively, unstable and stable.

We now show that the Fisher equation has a *traveling wave solution*. To do this, we assume that

$$u(x,t) = f(x-ct), \quad z = x - ct, \tag{5.4.38}$$

and substitute this into Equation (5.4.36) to obtain

$$-cf' = f'' + f - f^2, \tag{5.4.39}$$

or

$$f'' + cf' + f - f^2 = 0, \quad f' = \frac{df(z)}{dz} \tag{5.4.40}$$

Now for small values of $f(z)$, i.e.,

$$0 < f(z) \ll 1, \tag{5.4.41}$$

the linear approximation to Equation (5.4.40) is

$$f'' + cf' + f \simeq 0. \tag{5.4.42}$$

The corresponding characteristic equation is

$$r^2 + cr + 1 = 0, \tag{5.4.43}$$

if

$$f \sim e^{rz}. \tag{5.4.44}$$

Solving for the two values of r gives

$$r_\pm = \left(\frac{1}{2}\right)\left[-c \pm \sqrt{c^2 - 4}\right], \tag{5.4.45}$$

and, also the result

$$r_- < r_+ < 0. \tag{5.4.46}$$

Consequently, for small values of $f(z)$, which holds for z large and positive, $f(z)$ can be expressed as

$$z \text{ large } > 0 : f(z) \sim A e^{-|r_+|z}, \quad A > 0. \tag{5.4.47}$$

Likewise, for $f(z)$ near 1, we can write

$$f(z) = 1 - \epsilon(z), \quad 0 < \epsilon(z) \ll 1, \tag{5.4.48}$$

and therefore,
$$\epsilon'' + c\epsilon' - \epsilon = 0, \tag{5.4.49}$$

with the characteristic equation
$$s^2 + cs - 1 = 0, \quad \text{for } \epsilon(z) = e^{sz}. \tag{5.4.50}$$

Solving for s gives
$$s_\pm = \left(\frac{1}{2}\right)\left[-c \pm \sqrt{c^2 + 4}\right], \tag{5.4.51}$$

with
$$s_- < 0 < s_+. \tag{5.4.52}$$

Therefore,
$$z\text{large} < 0 : f(z) \sim 1 - Be^{|s_-|z}, \quad B > 0. \tag{5.4.53}$$

Comments

(i) A and B are unknown constants since the ODEs are linear.

(ii) The equilibrium solutions, $\bar{u}^{(1)}(x,t) = 0$ and $\bar{u}^{(2)}(x,t) = 1$, and the condition, $0 \le u(x,t) \le 1$, require that the speed of the traveling wave satisfies
$$c \ge 2. \tag{5.4.54}$$

(iii) Figure 5.3 sketches the trajectory in the (f, f') phase-plane for $c \ge 0$ and $0 < c < 2$.

(iv) In summary, the traveling wave solution, $u(x,t) = f(x - ct) = f(z)$, has the following features:

$$0 \le f(z) \le 1, \quad -\infty < z < +\infty; \tag{5.4.55}$$

$$\lim_{z \to -\infty} f(z) = f(-\infty) = 1, \quad \lim_{z \to +\infty} f(z) = f(+\infty) = 0; \tag{5.4.56}$$

$$f'(z) < 0, \quad -\infty < z < +\infty; \tag{5.4.57}$$

$$c \ge 0. \tag{5.4.58}$$

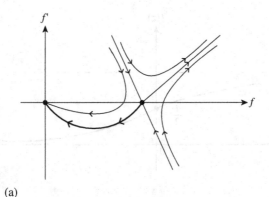

(a)

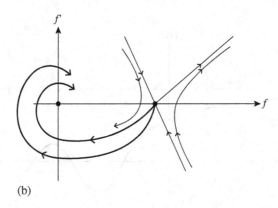

(b)

FIGURE 5.3 Phase-plane trajectories. (a) $c \geq 2$. (b) $0 < c < 2$.

(v) $f(z)$ is sketched in Figure 5.4 for the two cases, $c \geq 2$ and $0 < c < 2$.

(vi) An analytical, mostly ad hoc, approximation for the traveling wave solution of the Fisher equation is

$$f_{app}(z) = \frac{1}{1 + \exp(|r_+| z)}. \tag{5.4.59}$$

Note that it satisfies all of the conditions listed in Comment (iv) above. But, whether this expression is satisfactory or not depends on what the user needs and what they wish to accomplish.

5.4.4 Heat PDE

It is often stated that the heat PDE does not have traveling wave solutions, i.e., bounded solutions existing over the interval, $-\infty x <$

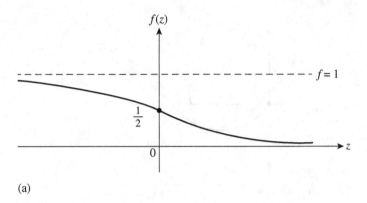

(a)

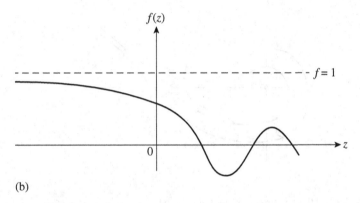

(b)

FIGURE 5.4 Traveling waves for the Fisher equation. (a) $c \geq 2$. (b) $0 < c < 2$.

$+\infty$, with bounded first and second derivatives. To see what this means, consider the heat equation

$$u_t = Du_{xx}, \quad u = u(x, t), \quad -\infty < x < +\infty, \quad (5.4.60)$$

and assume a solution of the form

$$u(x, t) = f(x - ct) = f(z), z = x - ct. \quad (5.4.61)$$

Using this in Equation (5.4.60) gives the second-order, linear ODE

$$-cf' = Df'', \quad f' = df/dz. \quad (5.4.62)$$

The general solution for this ODE is

$$f(z) = a + b \exp\left(-\frac{c}{D}\right)z, \quad (5.4.63)$$

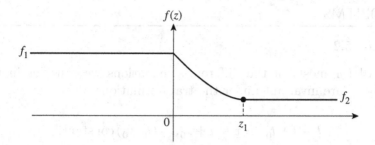

FIGURE 5.5 Piece-wise-continuous traveling wave solution to the heat equation. Definition given in Equation (5.4.65).

where (a, b) are arbitrary constants such that they are selected to satisfy the condition

$$f(z) \geq 0. \tag{5.4.64}$$

Inspection of Equation (5.4.63) shows that $f(z)$ is not bounded on the interval, $-\infty < z < +\infty$, but it does have derivatives of all orders.

It is important to point out that bounded, piecewise-continuous solutions can be constructed for the interval, $-\infty < z < +\infty$. Figure 5.5 provides an example of such a function. Its definition is given by the expression

$$f(z) = \begin{cases} f_1 > 0, & -\infty < z \leq 0; \\ f_1 \exp\left(-\dfrac{c}{D}\right) z, & 0 \leq z \leq z_1; \\ f_2, \, z \geq z_1, \end{cases} \tag{5.4.65}$$

where

$$f_1 > f_2 > 0, \quad z_1 = \left(\dfrac{D}{c}\right) \mathrm{Ln}\left(\dfrac{f_1}{f_2}\right). \tag{5.4.66}$$

Note that while $f(z)$ is continuous, it does not have a derivative at $z = 0$ or $z = z_1 > 0$.

A question to be asked is whether traveling wave solutions such as that indicated in Figure 5.5 correspond to the actual physical solution of some system that can be investigated in a laboratory? In the end, an answer must be provided by experiment rather than mathematics.

PROBLEMS

Section 5.2

1) All (or most) of the differential equations we consider in this book are invariant under the transformations

$$t \to t + t_0, \quad x \to x + x_0, \quad (t_0, x_0) \text{ constant}.$$

What are the implications of this result?

Section 5.3

2) Are there general rules for determining when an ODE or PDE is separable? If so, what are they? If not, why?

Section 5.4

3) As models for physical systems, what arguments would you give in support of having piece-wise-continuous functions as traveling wave solutions?

4) As models for physical systems, what arguments would you give in support of ruling out discontinuous functions as corresponding to the actual behavior of such systems?

NOTES AND REFERENCES

Section 5.1: For readers who wish to acquire a good background knowledge of the basic properties and related features of ODEs and PDEs, the following publications can be consulted

1. N. A. Kudryashov, Seven common errors in finding exact solutions of nonlinear differential equations, *Communications in Nonlinear Science and Numerical Simulation*, Vol. 14 (2010), 3507–3529.

2. J. D. Logan, *An Introduction to Nonlinear Partial Differential Equations* (Wiley, New York, 1994).

3. A. D. Polyanin and V. Zaitsev, *Handbook of Nonlinear Partial Differential Equations* (Chapman & Hall/CRC, Boca Raton, FL, 2004).

4. A. D. Polyanin and V. Zaitser, *Handbook of Exact Solutions for Ordinary Differential Equations*, 2nd Edition (Chapman & Hall/CRC, Boca Raton, FL, 2003).

Section 5.2: A broad discussion of symmetry and its role in the formulation and understanding of differential equations is provided in the following book

5. W. Miller, *Symmetry and Separation of Variables* (Cambridge University Press, New York, 1984).

Section 5.3 and 5.4: These two sections are based on my book

6. R. E. Mickens, *Mathematical Methods for the Natural and Engineering Sciences*, 2nd Edition (World Scientific, Singapore, 2017). See, in particular chapter 10.

Other useful references include

7. L. Debnath, *Nonlinear Partial Differential Equations* (Birkhäuser, Boston, 1997).

8. J. D. Murray, *Mathematical Biology* (Springer-Verlag, Berlin, 1989).

9. A. D. Polyanin and A. I. Zhurou, *Separation of Variables and Exact Solutions to Nonlinear PDE's* (Chapman and Hall/CRC, Boca Raton, FL, 2022).

10. Wikipedia: Separation of variables, https://en.wikipedia.org/wiki/ Separation - of - variables (Accessed July 4, 2024). Section 5.4: While being very selected techniques for obtaining exact solutions to both linear and nonlinear differential equations, a number of procedures have been created to calculate SOV and traveling wave solutions. Some recent references are

11. W. Malflient and W. Hereman, The tanh method:I. Exact solutions of nonlinear evolution and wave equations, *Physica Scripta*, Vol. 54 (1996), 563–568.

12. D. Bazeia, A. Das L. Losano, and A. Silva, A simple and direct method for generating traveling wave solutions for nonlinear waves, *Annuals of Physics*, Vol. 323 (2008), 1150–1167.

Note

13. (Authors Note, same as in ref. (12)): Traveling wave solutions of nonlinear differential equations, *Applied Mathematics Letters*, Vol. 23 (2010), 681–686.

14. G. W. Griffiths and W. E. Schiesser, *Traveling Wave Analysis of Partial Differential Equations* (Elsevier, Amsterdam, 2012).
15. R. Achouri, *Travelling Wave Solutions* (Dissertation, School of Mathematics, 2016). https://personalpages.manchester.ac, uk/staff/yanghong.huang/Projects/passprojects/travelwave. pdf (Accessed June 25, 2024).
16. S. W. Cho, H. J. Hwang, and H. Son, Traveling wave solutions of partial differential equations via neural networks, *Journal of Scientific Computing*, Vol. 89 (2021), Article 21.

For the adventurous, the following references provide insights into how separation of variables techniques and traveling wave methodologies are used to obtain solutions and insights for a broad range of problems at the cutting edge of research.

17. V. M, Villaba and E, I. Catala, Separation of variables and exact solution of the Klein-Gordon and Dirac equations in an open universe, see arXiv:gr-qc/0208017v1 (Accessed August 7, 2002).
18. A. Karbalaie, H. H. Muhammed, M. Shabaini, and M. M. Montazeri, Exact solution of partial differential equations using homoseparation of variables, *International Journal of Nonlinear Science*, Vol. 17 (2014), 84–90.
19. J. Manafian and M. Paknezhad, Analytical solutions for the Black–Scholes equation, *Applications and Applied Mathematics*, Vol, 12 (2017), 843–852.
20. A. F. Barannyk, T. A. Barannyk, and I. I. Yuryk, Exact solutions with generalized separation of variables in the nonlinear heat equation, *Ukrainian Mathematical Journal*, Vol. 74 (2022), 330–349.
21. R. Thomas and T. Bakkyaraj, Exact solutions of time-fractional differential-difference equations: Invariant subspace, partially invariant subspace, generalized separation of variables, *Computational and Applied Mathematics*, Vol. 43 (2024), Article 51.

CHAPTER 6

Introduction to Bifurcations

6.1 INTRODUCTION

To obtain a 'feel' for the subject of bifurcations, consider the differential equation

$$m\frac{d^2x}{dt^2} = -kx - \alpha\frac{dx}{dt}. \qquad (6.1.1)$$

This ordinary differential equation (ODE) provides a mathematical model for a damped harmonic oscillator of mass, m; spring constant, k; having a linear damping force characterized by the constant α(Figure 6.1). Experimentally, the appropriate initial conditions are

$$x(0) = x_0 > 0, \quad y(0) = y_0 = \frac{dx(0)}{dt} = 0. \qquad (6.1.2)$$

Note that, as currently written, this ODE has four associated parameters: m, k, α, x_0. However, we can rescale the ODE using

$$\begin{cases} x(t) = x_0 \bar{x}(\bar{t}), & t = T_1 \bar{t}, \\ T_1 = \sqrt{\frac{m}{k}}, & T_2 = \frac{m}{\alpha}, \quad 2\epsilon = \frac{T_1}{T_2}. \end{cases} \qquad (6.1.3)$$

In the weak or small damping case, we have

$$T_1 \ll T_2 \longrightarrow 0 < \epsilon \ll 1. \qquad (6.1.4)$$

This last restriction just means that the period of oscillation is small in comparison to the damping time. If we make these substitutions into Equation (6.1.1) and simplify the resulting expression, then we obtain

$$\frac{d^2\bar{x}}{d\bar{t}^2} + 2\epsilon\frac{d\bar{x}}{d\bar{t}} + \bar{x} = 0, \quad \bar{x}(0) = 1, \frac{d\bar{x}(0)}{d\bar{t}} = 0, \qquad (6.1.5)$$

or, dropping the bars

$$\frac{d^2x(t)}{dt^2} + 2\epsilon\frac{dx(t)}{dt} + x(t); \quad x(0) = 1, \quad \frac{dx(0)}{dt} = 0. \quad (6.1.6)$$

Now, if we rewrite this ODE as

$$\frac{dx}{dt} = y, \quad \frac{dy}{dt} = -x - 2\epsilon y; \quad x_0 = 1, \quad y_0 = 0, \quad (6.1.7)$$

then a 2-dim phase-plane analysis produces the results shown in Figure 6.2.

Observe that there are three cases to examine.

Case I: $0 < \epsilon \ll 1$

For this situation, the system oscillates with a damped amplitude. Also, we clearly see that

$$\lim_{t \to \infty}\begin{pmatrix} x(t) \\ y(t) \end{pmatrix} = \begin{pmatrix} 0 \\ 0 \end{pmatrix}, \quad (6.1.8)$$

in other words, the fixed-point, $(\bar{x}, \bar{y}) = (0, 0)$, is a stable spiral.

Case II: $\epsilon = 0$

For this case, the fixed-point, $(\bar{x}, \bar{y}) = (0, 0)$, is a center or has neutral stability.

Case III: $\epsilon < 0$

For this case, the fixed-point, $(\bar{x}, \bar{y}) = (0, 0)$ is an unstable spiral node and

$$\lim_{t \to \infty}\begin{pmatrix} x(t) \\ y(t) \end{pmatrix} = \begin{pmatrix} \infty \\ \infty \end{pmatrix}. \quad (6.1.9)$$

Comments

(i) It is important to note that for the 'physical ODE', Equation (6.1.1), the problem had four parameters, (m, k, α, x_0), but after rescaling the ODE, the number of parameters was reduced to one, namely, ϵ.

The critical lesson is that given an ODE or partial differential equation (PDE), we should always rescale both the dependent and independent variables to reduce the number of parameters.

Introduction to Bifurcations ■ 139

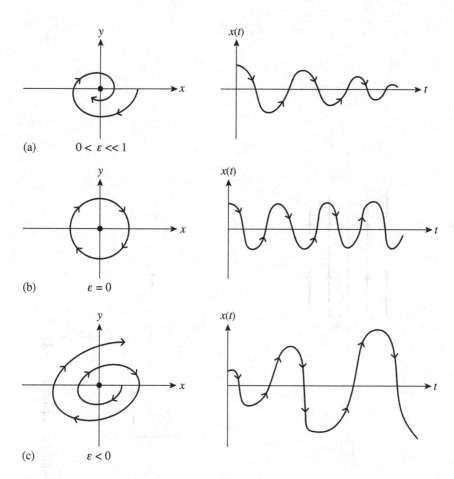

FIGURE 6.1 The damped harmonic oscillator for various epsilon values of the scaled damping coefficient ϵ.

(ii) For our particular example, the qualitative nature of the ODE's solutions depended on the value of the parameter ϵ, i.e.,

item $\epsilon > 0$: bounded, decreasing oscillatory solutions;
item $\epsilon = 0$: bounded, periodic solutions;
item $\epsilon < 0$: unbounded, increasing oscillatory solutions.

(iii) We will call ϵ a bifurcation parameter and $\epsilon = 0$, the bifurcation point for the rescaled damped harmonic oscillator equation, i.e., Equation (6.1.6).

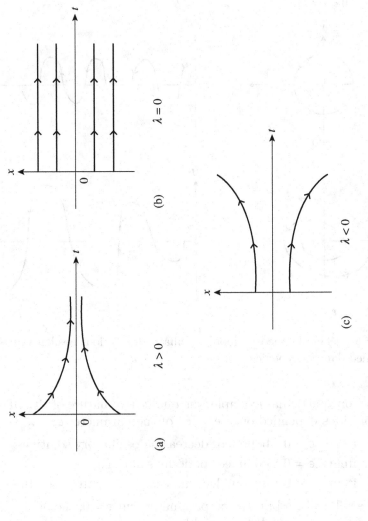

FIGURE 6.2 Qualitative behavior of the solutions for $dx/dt = -\lambda x$.

The remainder of this chapter will be devoted to specific examples of elementary equations that feature a select set of bifurcation phenomena. The only exception will be Section 6.2, where we present a working definition of bifurcation and discuss briefly why such a concept is relevant to the analysis of ODEs and PDEs that arise in the modeling process.

6.2 DEFINITION

Differential equations modeling physical systems also contain parameters corresponding to properties of the system, which in general are assumed to be either constant or to change slowly with respect to both time and other variables. For our purposes, we take the following as the definition of a bifurcation which depends on a single parameter, λ.

6.2.1 Bifurcation

Consider a differential equation

$$\frac{dx(t)}{dt} = f(x(t), t, \lambda), \qquad (6.2.1)$$

where $X(t)$ may be a scalar or vector function, and let $f(x, t, \lambda)$, scalar or function, depend on a single real parameter, λ. Then x will also depend on λ, i.e., if x is a solution of Equation (6.2.1), then

$$x = x(t, \lambda). \qquad (6.2.2)$$

A value, $\lambda = \lambda^*$, of the parameter, λ, will be called a *bifurcation point* of the differential equation, if the qualitative properties of its trajectories in phase-space change their fundamental character as λ passes through λ^*. The value, $\lambda = \lambda^*$, is called the bifurcation point with respect to λ.

This behavior was demonstrated in Section 6.1 with regard to the damped harmonic oscillator.

Differential equations are generally dependent on one or more parameters, the values of which can change to reflect the precise nature of the system to be modeled. Thus, even for the most elementary differential equations, we expect bifurcations to be present, as the following ODE demonstrates.

$$\frac{dx(t)}{dt} = -\lambda x(t). \qquad (6.2.3)$$

It should be clear that $\lambda = \lambda^* = 0$ is the bifurcation point, and that three classes of solutions exist, depending on whether

$$\lambda < 0, \quad \lambda = 0, \quad \lambda > 0. \qquad (6.2.4)$$

See Figure 6.2.

While we will only briefly cover several of the more elementary types of bifurcations, there does exist a broad class of articles, books, lecture notes and videos on the many forms that bifurcations can appear. In fact, new classifications of bifurcations regularly appear and the general subject is being actively investigated by many researchers. We provide below a short list of items that we have used to understand the subject of bifurcation.

(1) J. Hale and H. Kocak, *Dynamics and Bifurcations* (Springer-Verlag, New York, 1991).

(2) G. Iooss and D. D. Joseph, *Elementary Stability and Bifurcation Theory* (Springer-Verlag, New York, 1980).

(3) M. Hazewinkel, Bifurcation phenomena: A short introductory tutorial with examples, in M. Hazewinkel, R. Jurkovich, and J. H. P. Paelinck (editors) *Bifurcation Analysis* (Springer, Dordrecht, 1985).

(4) H. Kielhofer, *Bifurcation Theory: An Introduction with Applications to Partial Differential Equations* (Springer, Berlin, 2012).

(5) T. Ma and S. Wang, *Bifurcation Theory and Applications* (World Scientific, Singapore, 2005).

(6) S. Wiggins, *Global Bifurcations and Chaos* (Springer, New York, 2013).

6.3 EXAMPLES OF ELEMENTARY BIFURCATIONS

For 1-dim systems

$$\frac{dx}{dt} = f(x), \qquad (6.3.1)$$

there are essentially only three kinds of bifurcations that occur; they are

(1) saddle node
(2) transcritical
(3) pitchfork
   - supercritical
   - subcritical

In this section, we discuss briefly the saddle node, transcritical and supercritical pitchfork bifurcation and provide references to examinations of the subcritical pitchfork bifurcation.

6.3.1 Saddle-node Bifurcation

Consider the following ODE where r is real

$$\frac{dx}{dt} = r + x^2. \qquad (6.3.2)$$

This equation has two fixed-points or equilibrium solutions located at

$$\bar{x} = \pm\sqrt{-r}. \qquad (6.3.3)$$

There are three cases to consider:

Case 1: $r < 0$

For this situation, two real roots occur and the fixed-points are

$$\bar{x}_1 = -\sqrt{|r|}, \quad \bar{x}_2 = +\sqrt{|r|}, \qquad (6.3.4)$$

and the $x-t$ phase-plane, along with several typical solution behaviors are sketched in Figure 6.3. Note that the two fixed-points have the following stability properties.

$$\begin{cases} x(t) = \bar{x}_1 = -\sqrt{|r|} & : \text{ stable}, \\ x(t) = \bar{x}_2 = +\sqrt{|r|} & : \text{ unstable}, \\ r < 0. \end{cases} \qquad (6.3.5)$$

Case 2: $r = 0$

There is now a double root at $x = 0$ and the fixed-points are $\bar{x}_1 = \bar{x}_2 = 0$. The nature of these fixed-points means that $x = 0$ is semi-stable. See Figure 6.4.

144 ■ Introduction to Qualitative Methods for Differential Equations

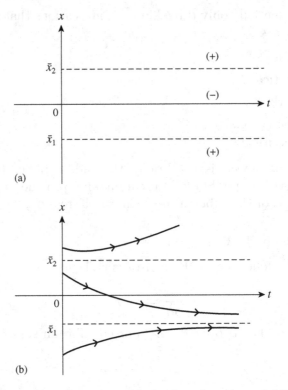

FIGURE 6.3 x–t plane, for $r < 0$, with the domains where dx/dt has definite signs. The equilibrium solutions are $x_1(t) = \bar{x}_1$ and $x_2(t) = \bar{x}_2$.

Case 3: $r > 0$

For this case, there is no (real) fixed-point and all solutions become unbounded.

Finally, Figure 6.5 is a sketch of the bifurcation diagram for Equation (6.3.2), i.e., the plot of $\bar{x}$ vs r. This case provides the general features of a so-called saddle-node bifurcation.

6.3.2 Transcritical Bifurcation

The prototypical example of a system exhibiting a transcritical bifurcation is the ODE

$$\frac{dx}{dt} = rx - x^2, \qquad (6.3.6)$$

Introduction to Bifurcations ■ 145

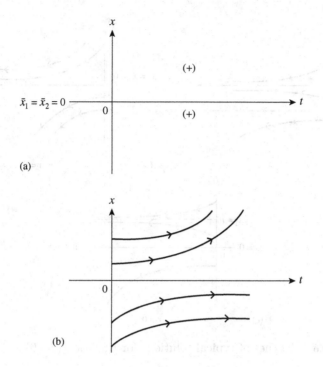

FIGURE 6.4 $x-t$ plane, for $r = 0$. The double fixed-point at $\bar{x}_1 = \bar{x}_2 = 0$ means that $x(t) = 0$ is semi-stable.

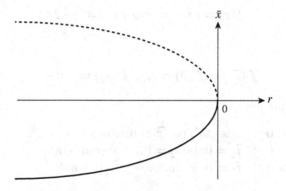

FIGURE 6.5 Bifurcation diagram for a saddle-node bifurcation, $r = \bar{x}^2$.

where r is real. For this case, there are two fixed-points

$$\bar{x}_1 = 0, \bar{x}_2 = r. \tag{6.3.7}$$

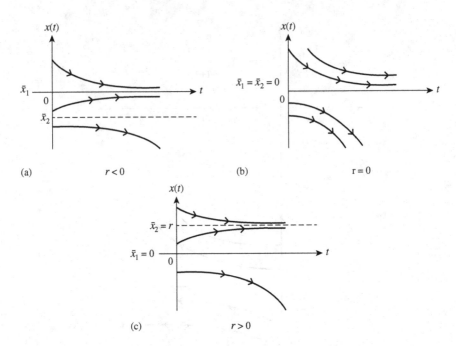

FIGURE 6.6 Sketches of typical solutions for Equation (6.3.6).

Since
$$f(x) = rx - x^2 \implies f'(x) = r - 2x \qquad (6.3.8)$$

and
$$f'(\bar{x}_1) = f'(0) = r, \quad f'(\bar{x}_2) = r = -r, \qquad (6.3.9)$$

it follows that for

$$\begin{cases} r < 0 \ : \ \bar{x}_1 \text{ is stable}, \ \bar{x}_2 \text{ is unstable}; \\ r = 0 \ : \ \bar{x}_1 = 0 \text{ and } \bar{x}_2 = 0, \text{ are semi stable}; \\ r > 0 \ : \ \bar{x}_1 = 0 \text{ is unstable}, \ \bar{x}_2 \text{ is stable}. \end{cases} \qquad (6.3.10)$$

See Figure 6.6 for typical $x - t$ sketches for various intervals of the parameter r. The bifurcation diagram for this case is given in Figure 6.7. The name associated with this bifurcation is that it is a *transcritical bifurcation*.

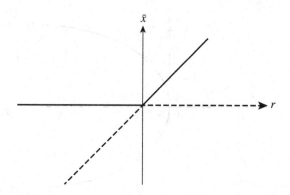

FIGURE 6.7 Bifurcation diagram for Equation (6.3.6) showing its transcritical bifurcation behavior.

6.3.3 Supercritical Pitchfork Bifurcation

This type of bifurcation is shown by the ODE

$$\frac{dx}{dt} = rx - x^{(3)}. \qquad (6.3.11)$$

Note that there are three fixed-points located at the positions

$$\bar{x}_1 = 0, \quad \bar{x}_2 = \sqrt{r}, \quad \bar{x}_3 = -\sqrt{r}. \qquad (6.3.12)$$

If we define

$$f(x) = rx - \bar{x}^3, \qquad (6.3.13)$$

then

$$f'(x) = r - 3x^2, \qquad (6.3.14)$$

and

$$f'(0) = r, \quad f'(\sqrt{r}) = -2r, \quad f'(-\sqrt{r}) = -2r. \qquad (6.3.15)$$

Consequently, combining these results with linear stability theory, it can be concluded that

(i) $r < 0$: Only the fixed-point $\bar{x}_1 = 0$ exists and it is stable.

(ii) $r = 0$: There is a triple zero at $x = 0$ and $\bar{x}_1 = \bar{x}_2 = \bar{x}_3 = 0$, and this fixed-point is stable.

(iii) $r > 0$: The fixed-point at $\bar{x}_1$ becomes unstable and two new fixed-points appear at $\bar{x} = \pm\sqrt{r}$.

This situation is called a supercritical pitchfork bifurcation and a representation of its diagram, i.e., $\bar{x}$ vs r is given in Figure 6.8.

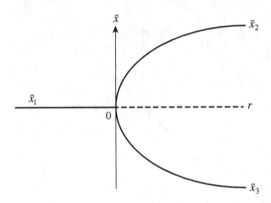

FIGURE 6.8 Bifurcation diagram for a supercritical pitchfork bifurcation for Equation (6.3.11).

6.3.4 Subcritical Pitchfork Bifurcation

This bifurcation appears in the ODE

$$\frac{dx}{dt} = rx + x^3 - x^5 \tag{6.3.16}$$

and can have as many as five fixed points. The bifurcation diagram, given in Figure 6.9, for this ODE shows the major features of the associated subcritical pitchfork bifurcation. The details of how this figure is constructed and its important properties can be found in the references

> S. Strogatz, *Non-linear Dynamics and Chaos: With Applications to Physics, Biology, Chemistry and Engineering* (Perseus Books, New York, 2000).

> S. Wiggins, *Introduction to Applied Non linear Dynamical Systems and Chaos* (Springer-Verlag, Berlin, 1990).

However, a quick summary goes as follows:

The fixed-points of Equation (6.3.16) are the solutions of

$$\bar{x}\left(r + \bar{x}^2 - \bar{x}^4\right) = 0. \tag{6.3.17}$$

The fixed-point, $\bar{x} = 0$, exists for all values of the parameter, r.

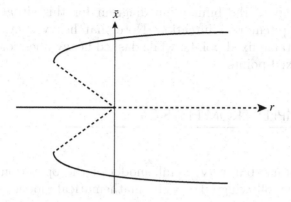

FIGURE 6.9 Diagram for the subcritical pitchfork bifurcation. Solid and dashed lines represent, respectively, stable and unstable fixed-points.

Also, the quartic equation can be solved for $\bar{x}^2$, which then can be solved for $\bar{x}$.

$$\bar{x}^4 - \bar{x}^2 - r = 0, \tag{6.3.18}$$

These four roots are

$$\bar{x} = (\pm)\sqrt{\left(\frac{1}{2}\right)\left(1 \pm \sqrt{1 + 4r}\right)}. \tag{6.3.19}$$

Inspection of Equation (6.3.19) shows that to have $\bar{x}$ real, then from the inner square root.

$$r \geq -\left(\frac{1}{4}\right). \tag{6.3.20}$$

Also, from the outer square root, $1 - \sqrt{1 + 4r}$ becomes negative for $r > 0$. Therefore, there are three intervals in r to consider and these intervals and the values of the associated fixed-points are given by the following expressions:

$r < -\left(\frac{1}{4}\right)$: $\bar{x} = 0$ (one fixed-point);

$-\left(\frac{1}{4}\right) < r < 0$: $\bar{x} = 0$, $\bar{x} = (\pm)\sqrt{\left(\frac{1}{2}\right)\left(1 \pm \sqrt{1 + 4r}\right)}$, five fixed-points;

$r > 0$: $\bar{x} = 0$, $\bar{x} = (\pm)\sqrt{\left(\frac{1}{2}\right)\left(1 + \sqrt{1 + 4r}\right)}$, three fixed-points.

Figure 6.9 gives the bifurcation diagram for this situation, i.e., a subcritical pitchfork bifurcation. Note that heavy continuous lines represent stable fixed-points, while dashed heavy lines correspond to unstable fixed-points.

6.4 EXAMPLES FROM PHYSICS

6.4.1 Lasers

An elementary, but very useful, model of the operation of a laser provides the following ODE as its mathematical model

$$\frac{dn(t)}{dt} = (GN_0 - k)n - (\alpha G)n^2, \tag{6.4.1}$$

where

n(t) : number of photons (light particles) in the laser light field;

N(t) : number of atoms in an excited state;

k : essentially the time that a photon spends in the active material of the laser before being loss through the end faces;

N0 : the initial number of excited atoms;

G : gain coefficient;

α : parameter coming from the assumption that

$$N(t) = N_0 - \alpha n(t), \quad (\alpha > 0). \tag{6.4.2}$$

If we define r as

$$r = GN_0 - k, \tag{6.4.3}$$

then Equation (6.4.1) takes the form

$$\frac{dn(t)}{dt} = rn(t) - (\alpha G)n(t)^2, \tag{6.4.4}$$

and, consequently, a transcritical bifurcation will exist for the laser model.

In physics, one usually constructs the bifurcation diagram from a plot of $\bar{n}$, the fixed-points, vs N_0. This is shown in Figure 6.10. Note that negative n and N_0 are physically meaning less.

The details of the theory of the laser are presented in

H. Haken, *Laser Theory* (Springer-Verlag, Berlin, 1983).

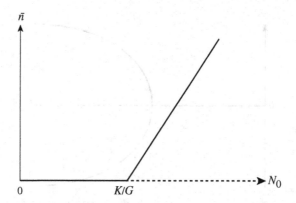

FIGURE 6.10 Bifurcation diagram for the laser equation. Note that for $0 < N_0 < K/G$, the laser device acts as a 'lamp.' For $N_0 > k/G$, lasering takes place.

6.4.2 Statistical Mechanics and Neural Networks

The following first-order, nonlinear ODE provides a model for many physical phenomena in the physics of magnetization and neural networks

$$\frac{dx}{dt} = -x + b\tanh(x), \qquad (6.4.5)$$

where b is generally a non-negative parameter, and the $\tanh(x)$ is defined to be

$$\tanh(x) = \frac{e^x - e^{-x}}{e^x + e^{-x}}. \qquad (6.4.6)$$

The fixed-points, $\bar{x}$, occur where the two curves

$$y = x, \quad y = b\tanh(x), \qquad (6.4.7)$$

intersect. With a rather minor effort, it can be determined that the bifurcation diagram, $\bar{x}$ vs b, has the form presented in Figure 6.11. Observe that for

(a) $0 < b \leq 1$: The fixed-point $\bar{x}_1 = 0$ is stable and there are no other fixed-points.

(b) $b > 1$: Three fixed-points exist; the one at $\bar{x}_1 = 0$ now becomes unstable, while the other two $\bar{x}_2$ and $\bar{x}_3$ are stable.

For case (b), which state the system settles into will depend on the initial condition for $x(t)$.

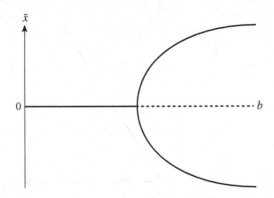

FIGURE 6.11 Bifurcation diagram for Equation (6.4.5) Heavy continuous lines are stable fixed-points. The heavy dashed line corresponds to an unstable fixed-point.

6.5 HOPF-BIFURCATIONS

6.5.1 Introduction

An important class of systems for which bifurcations may occur are those that have a certain type of oscillatory behavior. This section will be concerned with so-called Hopf-bifurcations in 2-dim dynamical systems, i.e., they can be modeled by differential equations of the form

$$\frac{dx}{dt} = P(x, y, \lambda), \frac{dy}{dt} = Q(x, y, \lambda), \tag{6.5.1}$$

where λ is a parameter.

The treatment of this topic is based on the presentation given in my book

R. E. Mickews, *Mathematical Methods for the Natural and Engineering Sciences*, 2nd Edition (World Scientific, Singapore, 2017). Sections 4.6.1 and 4.6.2.

To begin, assume that Equations (6.5.1) have an isolated fixed-point at $(\bar{x}(\lambda), \bar{y}(\lambda))$. We have explicitly indicated the dependence of the fixed-point on the parameter, λ. Now, for motions in a neighborhood of this fixed-point, we further assume that $x(t,\lambda)$ and $y(t,\lambda)$ have the structure

$$\begin{cases} x(t,\lambda) = \bar{x}(\lambda) + \alpha(t,\lambda), \\ y(t,\lambda) = \bar{y}(\lambda) + \beta(t,\lambda). \end{cases} \tag{6.5.2}$$

If Equation (6.5.2) is substituted into Equation (6.5.1) and only the linear terms are retained, then we have

$$\frac{d}{dt}\begin{pmatrix}\alpha\\\beta\end{pmatrix} = A\begin{pmatrix}\alpha\\\beta\end{pmatrix}, \qquad (6.5.3)$$

where the matric A is

$$A = \begin{pmatrix} \overline{\frac{\partial Q}{\partial x}} & \overline{\frac{\partial Q}{\partial y}} \\ \overline{\frac{\partial P}{\partial x}} & \overline{\frac{\partial P}{\partial y}} \end{pmatrix}, \qquad (6.5.4)$$

and where the bar over each term in the matrix implies that the element is evaluated at the fixed-point, $(\bar{x}(\lambda), \bar{y}(\lambda))$.

The eigenvalues of the matrix, A, are solutions of the equation

$$\det|A - rI| = 0 \qquad (6.5.5)$$

where I is the 2×2 unit matrix

$$[I = \begin{pmatrix} 1 & 0 \\ 0 & 1 \end{pmatrix}. \qquad (6.5.6)$$

Denote the two eigenvalues of A by $r_1(\lambda)$ and $r_2(\lambda)$, and assume that there exists a positive constant, δ, such that for, $|\lambda| < \delta$, $r_1(\lambda)$ and $r_2(\lambda)$ are complex-valued functions of λ, and also have a first derivative with respect to λ. In other words, we have

$$\begin{cases} r_1(\lambda) = R(\lambda) + iI(\lambda) \\ r_2(\lambda) = r_1^*(\lambda) \\ \frac{dr_1(\lambda)}{d\lambda} \text{ and } \frac{dr_2(\lambda)}{d\lambda}, \text{ exist,} \end{cases} \qquad (6.5.7)$$

where also $dR(\lambda)/d\lambda$ and $dI(\lambda)/d\lambda$ exist over the interval $|\lambda| < \delta$.

It should be indicated that we are taking the bifurcation point to be $\lambda = 0$.

Our version of the Hopf-bifurcation theorem takes the form:

6.5.2 Hopf-Bifurcation Theorem

Let the fixed-point, $(\bar{x}(\lambda), \bar{y}(\lambda))$, of Equation (6.5.1) be asymptotically stable for $\lambda < 0$ and unstable for $\lambda = 0$, Let $R(0) = 0$ and let

$$\frac{dR(\lambda)}{d\lambda}\Big|_{\lambda=0} > 0, \quad I(0) \neq 0. \qquad (6.5.8)$$

Under these conditions, for all sufficiently small $|\lambda|$, an isolated, closed trajectory exists for either $\lambda > 0$ or $\lambda < 0$. (This isolated, closed trajectory is called a limit-cycle.) In general, the stability of the limit-cycle is opposite to that of the fixed-point.

6.5.3 Fixed-Points and Closed Integral Curves

Let the system

$$\frac{dx}{dt} = P(x, y), \quad \frac{dy}{dt} = Q(x, y), \tag{6.5.9}$$

have a closed integral curve in the $x-y$ plane. Then this curve must enclose fixed-points whose 'indices' sum to $+1$:

Comments

(i) The index of the fixed-points that we have examined has the following indices

$$\begin{cases} \text{Fixed-Point} & \text{Index} \\ \text{node} & +1 \\ \text{saddle} & -1 \\ \text{center} & +1 \end{cases} \tag{6.5.10}$$

See Strogatz, Section 6.8, for a good introduction to this topic.

(ii) A consequence of this result is that every closed curve in the $x-y$ plane must contain at least one fixed-point.

(iii) Another implication of the above result is that if only one fixed-point is interior to a closed curve, then the fixed-point cannot be a saddle point.

See Figures 6.11 and 6.12 for illustrations of this result (Figure 6.13).

6.5.4 Two Limit-Cycle Oscillators

The van der Pol oscillator was one of the first nonlinear models that exhibited a limit-cycle. While it is usually written as

$$\frac{d^2x}{dt^2} + x = \lambda(1 - x^2)\frac{dx}{dt}, \tag{6.5.11}$$

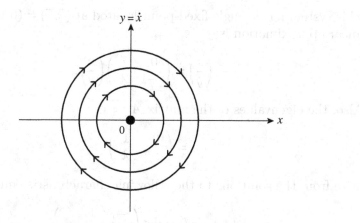

FIGURE 6.12 Phase-plane, $x-y$, for $\ddot{x}+x=0$. The fixed-point, $(\bar{x},\bar{y})=(0,0)$, is a center.

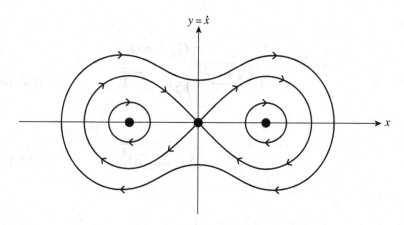

FIGURE 6.13 Phase-plane, $x-y$ for $\ddot{x}-x+x^3=0$. The fixed-points are $(\bar{x}_1,\bar{y}_1)=(0,0)$, a saddle point and the centers, $(\bar{x}_2,\bar{y}_2)=(1,0)$ and $(\bar{x}_3,\bar{y}_3)=(-1,0)$.

where λ is a parameter, we will use the form

$$\frac{d^2x}{dt^2}+x=(\lambda-x^2)\frac{dx}{dt}, \qquad (6.5.12)$$

in which system notation is

$$\frac{dx}{dt}=y, \quad \frac{dy}{dt}=-x+\lambda y-x^2 y. \qquad (6.5.13)$$

156 ■ Introduction to Qualitative Methods for Differential Equations

This system has a single fixed-point located at $(\bar{x}, \bar{y}) = (0,0)$, and its linear approximation is

$$\frac{d}{dt}\begin{pmatrix} x \\ y \end{pmatrix} = \begin{pmatrix} 0 & 1 \\ -1 & \lambda \end{pmatrix}\begin{pmatrix} x \\ y \end{pmatrix}. \tag{6.5.14}$$

Also, the eigenvalues of the matrix A

$$A = \begin{pmatrix} 0 & 1 \\ -1 & \lambda \end{pmatrix}, \tag{6.5.15}$$

arise from the solutions to the following characteristic equation

$$\det(A - rI) = \det\begin{pmatrix} -r & 1 \\ -1 & \lambda - r \end{pmatrix}$$
$$= r^2 - \lambda r + 1 = 0.$$

Thus,

$$r_1(\lambda) = r_2(\lambda)^*$$
$$= \frac{\lambda}{2} + \left(\frac{\sqrt{4-\lambda^2}}{2}\right)i, \quad i = \sqrt{-1}, \tag{6.5.16}$$

with

$$R(\lambda) = \frac{\lambda}{2}, \quad I(\lambda) = \frac{\sqrt{4-\lambda^2}}{2}. \tag{6.5.17}$$

Note that

$$R(0) = 0, \quad \frac{dR(0)}{d\lambda} = \frac{1}{2} > 0, \quad I(0) = 1. \tag{6.5.18}$$

Also, for $\lambda = 0$, Equation (6.5.12) becomes

$$\frac{d^2x}{dt^2} + x = -x^2\frac{dx}{dt}, \tag{6.5.19}$$

which is a linear harmonic oscillator with positive damping. Consequently, its solutions have the property

$$\lim_{t \to \infty}\begin{pmatrix} x(t) \\ y(t) \end{pmatrix} = \begin{pmatrix} 0 \\ 0 \end{pmatrix}, \tag{6.5.20}$$

which implies that the fixed-point, $(\bar{x}, \bar{y}) = (0,0)$, is stable.

Combining all of these results, it follows from the application of the Hopf-bifurcation theorem that the van der Pol oscillator, Equation (6.5.12), has a stable limit-cycle for small, positive values of the parameter λ.

A second example of a system having a limit-cycle is

$$\begin{cases} \frac{dx}{dt} = \lambda x - \omega y - (x^2 + y^2)(x + ay), \\ \frac{dy}{dt} = \omega x + \lambda y + (x^2 + y^2)(ax - y), \end{cases} \quad (6.5.21)$$

where the parameters, (a, ω) are held fixed and λ is the bifurcation parameter. The only fixed-point is at $(\bar{x}, \bar{y}) = (0, 0)$ and the linearization matrix is

$$A = \begin{pmatrix} \lambda & -\omega \\ \omega & \lambda \end{pmatrix},$$

with the eigenvalues solutions of the equation

$$(\lambda - r)^2 + \omega^2 = 0. \quad (6.5.22)$$

Therefore,

$$r_1(\lambda) = r_2^*(\lambda) \quad (6.5.23)$$

$$= \lambda + i\omega, \quad i = \sqrt{-1}, \quad (6.5.24)$$

and

$$R(\lambda) = \lambda, \quad I(\lambda) = \omega, \quad (6.5.25)$$

from which it follows that

$$R(0) = 0, \quad \frac{dR(0)}{d\lambda} = 1 > 0, \quad I(0) = \omega. \quad (6.5.26)$$

Based on the results contained within this last equation, we can conclude that the 2-dim system of ODEs, Equation (6.5.21), has small values of λ, a limit-cycle.

It should be indicated that Equations (6.5.21) can be exactly solved in terms of the elementary functions. The way that this can be accomplished is by carrying out the following steps:

(1) Transform from (x, y) to polar coordinates (r, θ), using

$$x(t) = r(t)\cos\theta(t), \quad y(t) = r(t)\sin\theta(t). \tag{6.5.27}$$

(2) Taking the derivatives of $x(t)$ and $y(t)$, substituting these results into Equation (6.5.21), and then solving for dr/dt and $d\theta/dt$, gives

$$\frac{dr}{dt} = (\lambda - r^2), \quad \frac{d\theta}{dt} = \omega + ar^2. \tag{6.5.28}$$

(3) Multiplying the first of these equations by r and defining $z(t)$ as

$$z(t) = r(t)^2, \tag{6.5.29}$$

we obtain

$$\frac{dz}{dt} = 2(\lambda - z)z, \tag{6.5.30}$$

and this ODE can be solved exactly.

(4) Substitution of $r(t) = \sqrt{z(t)}$ into the second expression in Equation (6.5.28) gives an ODE, which can be solved explicitly for $\theta(t)$.

Inspection of Equation (6.5.28) shows that the r-equation has fixed-points at

$$r_1(\lambda) = 0, \quad r_2(\lambda) = \sqrt{\lambda}, \quad r_3(\lambda) - \sqrt{\lambda}. \tag{6.5.31}$$

However, based on the fact that r is a polar coordinate, then it must satisfy the condition, $r \geq 0$. Therefore, only $r_1(\lambda) = 0$ and $r_2(\lambda) = +\sqrt{\lambda}$, have 'physical meaning.' Consequently, the two stationary solutions are

$$r_\lambda(\lambda) = 0 \; : \; x(t,\lambda) = 0, y(t,\lambda) = 0, \lambda \leq 0; \tag{6.5.32}$$

$$r_2(\lambda) = \sqrt{\lambda} \; : \; \begin{cases} x(t,\lambda) = \sqrt{\lambda}\cos[(\omega + a\lambda)t + \varphi_0], \\ y(t,\lambda) = \sqrt{\lambda}\sin[(\omega + a\lambda)t + \varphi_0], \end{cases} \tag{6.5.33}$$

where φ_0 is an arbitrary constant. Thus, in the $x-y$ phase-plane, the limit-cycle is a circle of radius $\sqrt{\lambda}$, since from Equations (6.5.32) and (6.5.33) an easy calculation gives

$$x^2 + y^2 = \lambda. \tag{6.5.34}$$

6.6 RESUMÉ

The results and discussions given in this chapter clearly demonstrate the utility of having some knowledge and understanding of bifurcations. It also shows that the application of the theory of bifurcations generally provides us with important qualitative properties of the solutions to differential equations in the absence of having exact solutions.

Finally, it should be obvious that while we have not discussed bifurcations in PDEs, they must occur. There is a rather large research literature on this topic and the following two publications provide a hint of what is being investigated and applications where these techniques can be used.

(1) H. B. Keller, Nonlinear bifurcation, *Journal of Differential Equations*, Vol. 7 (1970), 417–434.

(2) H. Kielh¨ofer, *Bifurcation Theory: An Introduction with Applications in Partial Differential Equations* (Springer, Berlin, 2012).

(3) H. Vecker, Continuation and bifurcation in nonlinear PDEs - Algorithms, applications, and experiments, *Jahresbericht der Deutschen Mathematiker - Vereinigung*, Vol. 124 (2022), 43–80.

(4) Y. A. Kuznetsov, *Elements of Applied Bifurcation Theory* (Springer, Berlin, 2023).

PROBLEMS

Section 6.3

1) Can the standard theory of bifurcations be applied to the following ODE?

$$\frac{d^2 x}{dt^2} + x^{\frac{1}{3}} = \left(\epsilon - x^2\right)\frac{dx}{dt}$$

If so, then apply it. If not explain why not? To aid in this task, consider the behavior of the trajectories in the $x-y$ phase-plane.

Section 6.4

2) The two examples considered in this section are first-approximations to rather complex phenomena. Look up the derivations of these equations and examine how they might be generalized.

Section 6.5

3) The Lewis oscillator equation can be written as

$$\frac{d^2x}{dt^2} + x = (\epsilon - |x|)\frac{dx}{dt}.$$

Does this equation have a Hopf-bifurcation?

COMMENTS AND REFERENCES

References to the relevant literature are given in the various sections.

CHAPTER 7

Applications

This chapter contains a number of applications, which illustrate the various techniques discussed in this book. Many of the applications use several different methods to obtain the desired approximations to the exact solutions of the differential equation under examination. We also demonstrate in several cases that even when exact solutions are available, their mathematical structures may be so complex (to the average researcher) that the additional use of qualitative methods is of great value to gaining an understanding of the most important features of the differential equations.

Some of the presented applicants are based directly on the work of the author or the author and his collaborators. In such cases, the writings in this book follow closely the published paper.

Finally, it should be pointed out that the main goal of this chapter is to indicate the very broad applicability of qualitative methods in the natural and engineering sciences.

7.1 ESTIMATION OF $Y(0)$ FOR A BOUNDARY-VALUE PROBLEM

Consider the following boundary-value problem

$$y''(z) = y(z)[y(z) - zy'(z)], 0 < z < \infty, \qquad (7.1.1)$$

$$y'(0) = -\sqrt{3}, y(\infty) = 0, \qquad (7.1.2)$$

where $y' = dy/dz$, with the requirements

$$y(z) > 0, \quad y'(z) < 0, 0 < z < \infty. \qquad (7.1.3)$$

DOI: 10.1201/9781003422419-8

The goals are first to determine $y(0)$ and then calculate $y(z)$. To date, no exact result has been found to accommodate either task. Consequently, we will construct an analytical approximation for $y(z)$ and use this to estimate $y(0)$.

The derivations of Equations (7.1.1) and (7.1.2) are provided in the book of Dressen [1]. The article by Mickens and Wilkins [2] provides a summary of this derivation and is the basis of the work to be presented in this section. It should be indicated that these equations follow from a similarity-based solution of the partial differential equation (PDE)

$$uu_t = u_{xx}, \quad x > 0, t > 0, \quad u = u(x,t), \quad (7.1.4)$$

with

$$u(x,t) > 0, \quad 0 < x < \infty, \quad 0 < t < \infty, \quad (7.1.5)$$

with the following initial- and boundary-value conditions:

$$u(x,0) = 0, x > 0, \quad (7.1.6a)$$

$$u(\infty, t) = 0, t > 0, \quad (7.1.6b)$$

$$u_x(0,t) = -1, \quad (7.1.6c)$$

7.1.1 Properties of $y(z)$

While an exact analytical solution to Equations (7.1.1) and (7.1.2) does not exist in terms of a finite combination of the elementary functions, it is still possible to determine the major features of these solutions.

(i) An exact, nontrivial solution is

$$y(z) = 0. \quad (7.1.7)$$

(ii) Another exact solution can be found by assuming that $y(z)$ takes the form

$$y(z) = Az^\alpha, \quad (7.1.8)$$

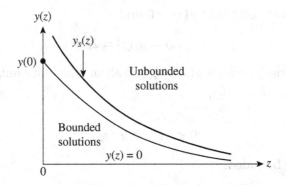

FIGURE 7.1 Bounded solutions lie below the $y_s(z)$ vs z curve, while the unbounded solutions are above this curve.

(A, α) are to be determined. Substituting this ansatz into the ordinary differential equation (ODE) and simplifying the resulting expression gives

$$\alpha(\alpha - 1) z^{\alpha-2} = A(1-\alpha) z^{2\alpha}, \qquad (7.1.9)$$

Consistency requires or

$$\alpha - 2 = 2\alpha, \quad \alpha(\alpha - 1) = A(1-\alpha), \qquad (7.1.10)$$

$$\alpha = -2, \quad A = 2, \qquad (7.1.11)$$

$$y(z) = y_s(z) = \frac{2}{z^2}, \qquad (7.1.12)$$

is an exact solution, which is singular in the sense that

$$y_s(0) = +\infty, \quad y'_s(0) = -\infty. \qquad (7.1.13)$$

Figure 7.1 provides a representation of the $y-z$ plane. Note that $y_s(z)$ is the boundary between the bounded and non-bounded solutions.

(iii) Inspection of Figure 7.1 shows that there exist three types of solutions:

(1) The exact solutions, $y(z) = 0$ and
$$y(z) = y_s(z) = 2/z^2.$$
(2) Solutions for which $y(z) > y_s(z)$; all of which are unbounded.
(3) Solutions for which
$$0 < y(z) < y_s(z), \qquad (7.1.14)$$
with all bounded.

Note that in more detail, the last equation is
$$0 < y(z) < \frac{2}{z^2}, \quad 0 < z < \infty. \qquad (7.1.15)$$

(iv) From the ODE, it follows that at $z = 0$, we have
$$y''(0) = y(0)^2. \qquad (7.1.16)$$

(v) If the ODE is differentiated twice and the results of Equation (7.1.3) are used, then it follows that
$$y''(z) > 0, y'''(z) < 0, y^{(4)}(z) > 0, 0 < z < \infty. \qquad (7.1.17)$$

The higher derivatives do not seem to have any particular pattern of sign.

(vi) The following sum-rule holds
$$\int_0^\infty y(z)^2 \, dz = \frac{2}{\sqrt{3}}. \qquad (7.1.18)$$

This result is a consequence of the following arguments. The term $zy(z)y'(z)$ can be rewritten as
$$zyy' = \left(\frac{1}{2}\right)(zy^2)' - \left(\frac{1}{2}\right)y^2. \qquad (7.1.19)$$

Therefore, the ODE takes the form
$$y'' = \left(\frac{3}{2}\right)y^2 - \left(\frac{1}{2}\right)(zy^2)'. \qquad (7.1.20)$$

Now integrate both sides from $z = 0$ to $z = \infty$ and obtain

$$y'(\infty) - y'(0) = \left(\frac{3}{2}\right)\int_0^\infty y(z)^2 dz - \left(\frac{1}{2}\right)(zy^2)\Big|_0^\infty. \quad (7.1.21)$$

Since

$$\begin{cases} y'(\infty) = 0, \quad y'(0) = -\sqrt{3}, \\ \lim_{z \to 0}(zy^2) = 0, \\ \lim_{z \to \infty}(zy^2) = 0, \end{cases} \quad (7.1.22)$$

the result in Equation (7.1.18) follows.

7.1.2 Approximation to $y(z)$

Since the exact solution, $y(z)$, is not known, we will 'assume' that a good, reasonable approximation is provided by the representation

$$y_a(z) = \frac{A}{1 + Bz + Cz^2}, \quad (7.1.23)$$

where the 'a' indicates the approximate nature of $y_a(z)$ and the parameters (A, B, C) are positive. Under these assumptions, we have

$$y_a(z) > 0, \quad y'_a(z) < 0, \quad 0 \le z < \infty. \quad (7.1.24)$$

Also, note that the inequalities stated in Equation (7.1.17) hold.

The parameter A can be determined by evaluating $y_a(0)$, i.e., setting $z = 0$ gives

$$A = y_a(0) = y(0), \quad (7.1.25)$$

where we have assumed that $y_a(0)$ takes the exact value $y(0)$, which is currently unknown.

If we take the derivative of $y_a(z)$ and let $z = 0$, we obtain

$$B = \frac{\sqrt{3}}{y(0)}, \quad (7.1.26)$$

where again we are taking this evaluation to be equal to the exact value given in Equation (7.1.2).

Taking into consideration the asymptotics of Equation (7.1.23), i.e.,

$$y_a(z) = \frac{2}{z^2} - O\left(\frac{1}{z^3}\right), \qquad (7.1.27)$$

where we indicate that the correction term is negative in sign, we determine C to be

$$C = \frac{y(0)}{2}, \qquad (7.1.28)$$

If we now put all these results together, then $y_a(z)$ is

$$y_a(z) = \frac{y(0)}{1 + \left[\frac{\sqrt{3}}{y(0)}\right]z + \left[\frac{y(0)}{2}\right]z^2}. \qquad (7.1.29)$$

Examination of this $y_a(z)$ shows that the right-hand side depends on the unknown, $y(0)$, which we had to determine. And, the way to do so is by substituting $y_a(z)$ for $y(z)$ in the sum-rule given by Equation (7.1.18), i.e.,

$$\int_0^\infty y_a(z)^2 \, dz = \frac{2}{\sqrt{3}}. \qquad (7.1.30)$$

With $y_a(z)$, given above, the left-hand side of the sum-role will be a function of $y(0)$, i.e.,

$$\int_0^\infty y_a(z)^2 \, dz = F(y(0)), \qquad (7.1.31)$$

where $F(y(0))$ is a very complication function of $y(0)$. However, another way of estimating $y(0)$ is to substitute $y_a(z)$ into the differential equation and setting $z = 0$. After a rather long, but straightforward set of manipulations, we find that $y(0) = y_0$ satisfies a cubic equation

$$y_0^3 + y_0^2 - 6 = 0, \quad y_0 = y(0). \qquad (7.1.32)$$

It is rather easy to show that this equation has one, real, positive root and two complex-conjugate root. The real root is

$$y_0 = y(0) = 1.537656174.... \qquad (7.1.33)$$

This is to be compared with Logan's value of

$$\text{Logan}: y(0) = 1.5111, \tag{7.1.34}$$

using a numerical method to 'solve' the ODE. The absolute and percentage errors between these two estimations of $y(0)$ are

$$Absolute\ Error = |(1.5377 - 1.5111| = 0.0266,$$
$$\%Error = |1.5377 - 1.51111.5111| \cdot (100) = 1.76\%.$$

Finally, an easy calculation provides the following asymptotics for $y_a(z)$

$$y_a(z) = \frac{2}{z^2} - \left[\frac{4\sqrt{3}}{y(0)^2}\right]\left(\frac{1}{z^3}\right) + O\left(\frac{1}{z^4}\right). \tag{7.1.35}$$

This result is consistent with the inequality

$$0 \leq y_a(z) < y_s(z), \quad 0 < z < \infty, \tag{7.1.36}$$

which shows that the approximate solutions lie below the singular solution.

7.1.3 Resume

Our task was to construct an approximation to the initial-value, boundary-value problem given by Equations (7.1.1) and (7.1.2), and also estimate the value, $y(0)$. This was done using a simple rational expression for the approximate solution, $y_a(z)$. Using this ansatz, we were able to estimate $y(0)$ as $y_a(0) = 1.537$.

The methodology presented can be easily generalized to other rational approximations for $y(z)$. In particular, the form

$$y_a(z) = \frac{P_N(z)}{Q_{N+2}(z)}, \tag{7.1.37}$$

is one possibility, where $P(z)$ and $Q(z)$ are polynomials, respectively, of degree N and $N+2$.

Finally, it should be noted that our analysis of this problem uses several different qualitative methods to reach a valid approximation, $y_a(z)$, to the solution $y(x)$.

7.2 THOMAS–FERMI EQUATION (TFE)

The Thomas–Fermi differential (TFE) equation is

$$\frac{d^2 y(x)}{dx^2} = \frac{y(x)^{(3/2)}}{x^{\frac{1}{2}}}, \qquad (7.2.1)$$

subject to the boundary conditions

$$y(0) = 1, \quad y(\infty) = 0. \qquad (7.2.2)$$

Historically, the TFE played a fundamental role in calculating the structure of atoms and still is a subject of interest in applied mathematics.

7.2.1 Exact Results

While the TFE cannot be solved analytically in terms of a finite combination of elementary functions, a number of properties of its solution, $y(x)$, are known and can be proved. The paper of Hille provides proofs for some of the statements to follow.

(i) The TFE has an exact solution

$$y(x) = y_s(x) = \frac{144}{x^3}. \qquad (7.2.3)$$

Note that this solution, $y_s(x)$, is a singular solution since it contains no arbitrary constants and is not a special case of the (unknown) general solution. Consequently, we expect that $y_s(x)$ is an 'attractor' in the sense that given a solution $y(x)$, it follows that

$$\lim_{x \to \infty} (x^3 y(x)) = 144. \qquad (7.2.4)$$

(ii) The curve, $y = y_s(x)$, separates the bounded and unbounded solutions of the TFE. See Figure 7.2 for the flow-space of the solutions.

Since $y(x) = 0$ is a solution, all solutions below the $y = y_s(x)$ curve are bounded and all solutions above it are unbounded. The bounded solutions all have a finite value of $y(0)$, while the unbounded solutions have $y(0) = +\infty$, i.e.,

$$\begin{cases} \text{bounded solutions:} 0 < y(0) < \infty, \\ \text{unbounded solutions:} y(0) = \infty. \end{cases} \qquad (7.2.5)$$

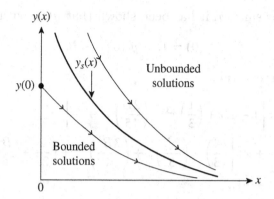

FIGURE 7.2 Trajectory flow for the Thomas–Fermi equation. Bounded solutions lie below the curve of the singular solution, $y = y_s(x)$, while unbounded solutions are above it.

(iii) The restrictions from atomic structure physics require that the following conditions hold

$$\begin{cases} y(x) > 0, y'(x) < 0, 0 \le x < \infty, \\ y(0) = 1, y(\infty) = 0. \end{cases} \quad (7.2.6)$$

Since $y''(x) > 0$, for $0 < x < \infty$, then $y(x)$ is concave upward as shown in Figure 7.2.

(iv) From Equations (7.2.1) and (7.2.2), we have

$$y''(x) = O\left(\frac{1}{\sqrt{x}}\right), \quad \text{as } x \to 0^+. \quad (7.2.7)$$

(v) For the bounded solutions of the TFE, the following conditions hold

$$\begin{cases} 0 < y(x) < y(0), \\ y'(0) < y'(x) < 0, \quad 0 < x < \infty, \\ y''(x) > 0, \quad 0 < x < \infty. \end{cases} \quad (7.2.8)$$

(vi) Arguments by Fernández demonstrate that $y'(0)$ can be calculated to essentially any decimal place of accuracy. Defining B as

$$y'(0) = -B < 0, \quad (7.2.9)$$

he obtains

$$B = 1.588071022611375.... \quad (7.2.10)$$

(vii) Also, for small x, it has been shown that $y(x)$, with

$$y(0) = 1, \quad y'(0) = -B, \qquad (7.2.11)$$

has the representation

$$y(x) = \left[1 - Bx + \left(\frac{1}{3}\right)x^3 - \left(\frac{2B}{15}\right)x^4 + \cdots\right]$$
$$+ x^{\frac{3}{2}}\left[\left(\frac{4}{3}\right) - \left(\frac{2B}{5}\right)x + \left(\frac{3B^2}{70}\right)x^2 + \left(\frac{2}{27} + \frac{B^3}{252}\right)x^3\right.$$
$$\left.+ \cdots\right].. \qquad (7.2.12)$$

This interesting result is found in the article: E. B. Baker, The application of the Fermi-Thomas statistical model to the calculation of potential distribution in positive ions, *Physical Review*, Vol. 16 (1930), 630–647.

(viii') The following sum-rules have been derived by Mickens and Herron:

$$\int_0^\infty \sqrt{x}\, y(x)^{\frac{3}{2}}\, dx = 1, \qquad (7.2.13)$$

$$\int_0^\infty y(x)\, dx = \left(\frac{1}{2}\right)\int_0^\infty x^{\frac{3}{2}} y(x)\, dx, \qquad (7.2.14)$$

$$B = \int_0^\infty \left[\left(\frac{dy}{dx}\right)^2 + \frac{y^{\frac{5}{2}}}{\sqrt{x}}\right] dx, \qquad (7.2.15)$$

$$B = \int_0^\infty \frac{y^{\frac{3}{2}}}{\sqrt{x}}\, dx. \qquad (7.2.16)$$

As an illustration to how these sum-rules are derived, let us do so for the one given in Equation (7.2.15).

First, multiply both sides of the TFE by $y(x)$ to obtain

$$yy'' = \frac{y^{\frac{5}{2}}}{\sqrt{x}}. \qquad (7.2.17)$$

Next, integrate this expression by parts and make evaluations at $x = 0$ and $x = \infty$; doing this gives

$$y(\infty) y'(\infty) - y(0) y'(0) - \int_0^\infty (y')^2 dx$$

$$= \int_0^\infty \frac{y^{\frac{5}{2}}}{\sqrt{x}} dx. \qquad (7.2.18)$$

Since

$$y(\infty) = 0, \quad y(0) = 1, \quad y'(\infty) = 0, \qquad (7.2.19)$$

then a rearrangement of terms gives the sum-rule expressed in Equation (7.2.15).

(ix) To indicate how the expansion given in Equation (7.2.12) can be obtained, the following procedure was 'created' by Mickens. For small x, the TFE can be approximated by

$$y''(x) \simeq \frac{1}{\sqrt{x}}, \qquad (7.2.20)$$

since $y(0) = 1$. If we integrate this ODE twice and use the initial conditions

$$y(0) = 1, \quad y'(0) = -B, \qquad (7.2.21)$$

then the following result is obtained

$$y(x) \simeq 1 - Bx + \left(\frac{4}{3}\right) x^{3/2}. \qquad (7.2.22)$$

Note that this corresponds to the first three lowest power terms in Equation (7.2.12).

An iteration scheme to calculate higher-order terms (taking care to be 'careful') is

$$y_{k+1}''(x) = \frac{[y_k(x)]^{\frac{3}{2}}}{\sqrt{x}}, \quad k = (0, 1, 2, \ldots), \qquad (7.2.23)$$

$$y_k(0) = 1, \quad y_k'(0) = -B, \qquad (7.2.24)$$

$$y_0(x) = 1. \qquad (7.2.25)$$

Note that $y_1(x)$ is the expression of Equation (7.2.22).

To obtain $y_2(x)$, we use

$$y_2''(x) = \frac{[y_1(x)]^{-3/2}}{\sqrt{x}}$$

$$= \frac{\left[1 - Bx + \left(\frac{4}{3}\right)x^{\frac{3}{2}}\right]^{\frac{3}{2}}}{\sqrt{x}}, \qquad (7.2.26)$$

and expand the bracketed expression to get

$$\left[1 - Bx + \left(\frac{4}{3}\right)x^{\frac{3}{2}}\right]^{\frac{3}{2}} = 1 - \left(\frac{3B}{2}\right)x + \cdots, \qquad (7.2.27)$$

where only the first two terms are kept. Therefore, with this, we must solve the following differential equation

$$y_2''(x) = \frac{1}{\sqrt{x}} - \left(\frac{3B}{2}\right)\sqrt{x}. \qquad (7.2.28)$$

Integrating this equation twice and imposing the initial conditions

$$y_2(0) = 1, \quad y_2'(0) = -B, \qquad (7.2.29)$$

we obtain

$$y_2(x) = 1 - Bx + \left(\frac{4}{3}\right)x^{\frac{3}{2}} - \left(\frac{2B}{5}\right)x^{\frac{5}{2}}, \qquad (7.2.30)$$

and this reproduces the first four lowest power terms of Equation (7.2.12).

7.2.2 Approximate Solutions

Two possible rational ansatzes for approximate solutions to the TFE are

$$y_a'(x) = \frac{1}{1 + Bx + \left(\frac{1}{144}\right)x^3}, \qquad (7.2.31)$$

$$y_a^2(x) = \frac{1}{1 + B \times - \left(\frac{4}{3}\right) x^{\frac{3}{3}} + Cx^2 + \left(\frac{1}{144}\right) x^3}, \quad (7.2.32)$$

where

$$C = \frac{B^2}{2} = 1.260985. \quad (7.2.33)$$

These mathematical expressions were selected to be dynamically consistent with the following properties of the TFE and/or its solutions:

1. $y(0) = 1, \quad y'(0) = -B;$ (7.2.34)

2. $0 \leq y(x) \leq 1, \quad 0 \geq x < \infty;$ (7.2.35)

3. $-B \leq y'(x) < 0, \quad 0 \leq x < \infty;$ (7.2.36)

4. $\lim_{x \to \infty} (x^3 y(x)) = 144;$ (7.2.37)

5. The approximations, taken to be rational in $\sqrt{x}$, should have one or the other forms for small x,

$$y_a(x) = \begin{cases} 1 - Bx, \\ 1 - Bx + \left(\frac{4}{3}\right) x^{\frac{3}{2}}. \end{cases} \quad (7.2.38)$$

7.2.3 Discussion

The qualitative features of the two rational approximations, $y_a^{(1)}(x)$ and $y_a^2(x)$, are given in Figure 7.3. Note that the solutions intersect at $x \simeq 1$. Observe that for $x > x_0$, $y_a^{(1)}(x)$ is greater than $y_a^2(x)$, which then appears to decrease much faster than $y_a^{(1)}(x)$. The paper of Mickens and Herron also provides a table giving numerical comparison of the values of $y_a^{(1)}(x)$ and $y_a^2(x)$ over the interval, $0 \leq x \leq 40$, along with the values obtained from numerical integration of the Thomas–Fermi ODE.

The reader should clearly understand that the TFE is itself an approximation to an (unknown) not fully characterized equation modeling atomic phenomena.

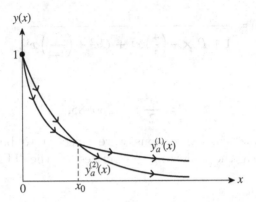

FIGURE 7.3 Schematic drawing of the approximations to solution of the Thomas–Fermi equation. The two solutions intersect at $x_0 \approx 1$.

Therefore, the actual, in use, value of this work and that of others in constructing approximate solutions for the TFE will be dependent on the weeds of the users of these solutions. What might 'work' in one set of circumstances might not prove valid if these conditions change.

A consistency check on the two approximate solutions is to use them in Equation (7.2.15) to calculate the value of $y'(0) = B$. A numerical integration gives

$$B_1 = 1.584744, \quad B_2 = 1.592931, \qquad (7.2.39)$$

where B_i is determined from the use of $y_a^{(i)}(x)$, for $i = (1, 2)$. The fractional percentage errors are

$$E_1 = \left(\frac{B - B_1}{B}\right) \cdot 100 = 0.21 \backslash \%, \qquad (7.2.40)$$

$$E_2 = \left(\frac{B - B_2}{B}\right) \cdot 100 = -0.27 \backslash \%. \qquad (7.2.41)$$

Thus, with respect to the use of a sum-rule to 'calculate' B, the two rational approximations give essentially the same result.

7.3 TRAVELING-WAVE FRONT BEHAVIOR FOR A PDE HAVING SQUARE-ROOT DYNAMICS

An important class of nonlinear PDEs is that associated with reaction–diffusion–advection phenomena. For one-space dimension, these equations take the form

$$u_t \pm g(u) u_x = D u_{xx} + f(u), \quad u = u(x, t). \tag{7.3.1}$$

Each term incorporates particular physical effects:

$$\begin{cases} u_t & : \text{ system evolution}, \\ g(u) u_x & : \text{ (nonlinear) advection}, \\ D u_{xx} & : \text{ diffusion} (D = \text{constant}), \\ f(u) & : \text{ reaction}. \end{cases}$$

Note that more complex PDEs can be formulated. For example, the abovementioned is assumed to be independent of u; but if this is not the case, then the diffusion term takes the form:

$$\text{Diffusion:} \frac{\partial}{\partial x}\left(D(u)\frac{\partial u}{\partial x}\right). \tag{7.3.2}$$

Also, for many physical systems, the reaction term has the following properties:

$$\begin{cases} f(u) > 0, & 0 < u < u^*, \\ f(0) = 0, & f(u^*) = 0, \end{cases} \tag{7.3.3}$$

and $f(u)$ takes the general shape indicated in Figure 7.4. Another important point is that for many physical systems, $u(x, t)$ is required to be non-negative. For these cases, $u(x, t)$ might represent a population number or a density. Further, $u(x, t)$ may be restricted to satisfy the positivity condition:

$$0 \le u(x, t) \le u^*. \tag{7.3.4}$$

An excellent summary and extensive discussion of reaction–diffusion–advection PDEs is given in the book by L. Debnath, *Nonlinear Partial Differential Equations for Scientists and Engineers* (Birkhäuser, Boston, 1997).

The main purpose of this section is to analyze the wave front behavior of the traveling wave solutions for a reaction–diffusion

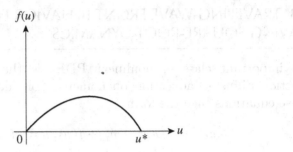

FIGURE 7.4 Typical form for reaction term, $f(u)$.

PDE having square-root dynamics. The particular equation to be examined is

$$u_t = Du_{xx} + \lambda_1\sqrt{u} - \lambda_2 u, \tag{7.3.5}$$

where the parameters $(D, \lambda_1, \lambda_2)$ are positive. Using scaling of the independent and dependent variables, we will show that Equation (7.3.5) can be rewritten to the form

$$\frac{\partial u}{\partial t} = \frac{\partial^2 u}{\partial x^2} + \sqrt{u} - u \tag{7.3.6}$$

in the scaled variables. Note that this PDE has two fixed-points or constant solutions; they are

$$u^{(1)}(x, t) = 0, \quad u^2(x, t) = 1. \tag{7.3.7}$$

A traveling wave (TW) solution of Equation (7.3.6) is a solution that takes the form

$$U(x, t) = f(x - ct)$$
$$= f(z), z = x - ct, \tag{7.3.8}$$

where C is the velocity of the TW. Note that a prior C is unknown and in many instances can either be explicitly calculated or have bounds placed on its possible values.

Further, to be a TW solution, $f(z)$ has the following properties:

$$\lim_{z \to -\infty} f(z) = 1, \quad \lim_{z \to +\infty} f(z) = 0, \tag{7.3.9}$$

$$\frac{df(z)}{dz} < 0, \quad -\infty < z < +\infty. \tag{7.3.10}$$

Since Equations (7.3.5) or (7.3.6) cannot be solved explicitly, this also means that the differential equation for $f(z)$ has this feature. However, our interest is in examining the properties of $f(z)$ when $f(z)$ is small, i.e., the values of z such that

$$0 \le f(z) << 1. \qquad (7.3.11)$$

The resulting functional form of $f(z)$ is called the *wave front behavior*.

7.3.1 Variable Scaling

We now show how Equation (7.3.6) can be from Equation (7.3.5) by means of a rescaling of the variables (u, t, x).

Let (T, L, W) be, respectively, the time, length and dependent variable u scales, i.e.,

$$t = T\bar{t}, \qquad x = L\bar{x}, \qquad u = W\bar{u}. \qquad (7.3.12)$$

Note that $(\bar{u}, \bar{t}, \bar{x})$ are the scaled, dimensionless new variables. Substitution of the results in Equation (7.3.12) into Equation (7.3.5) and rearranging terms give the expression

$$\frac{\partial \bar{u}}{\partial \bar{t}} = \left(\frac{TD}{L^2}\right) \frac{\partial^2 \bar{u}}{\partial \bar{x}^2} + \left(\frac{T\lambda_1}{\sqrt{W}}\right)\sqrt{\bar{u}} - (T\lambda_2)\bar{u}. \qquad (7.3.13)$$

If we require

$$\frac{TD}{L^2} = 1, \qquad \frac{T\lambda_1}{\sqrt{W}} = 1, \qquad T\lambda_2 = 1, \qquad (7.3.14)$$

and solve for the scales, then

$$T = \frac{1}{\lambda_2}, \qquad L = \sqrt{\frac{D}{\lambda_2}}, \qquad W = \left(\frac{\lambda_1}{\lambda_2}\right)^2. \qquad (7.3.15)$$

If the bars are now dropped over the scaled variables, then the PDE in Equation (7.3.13) becomes

$$\frac{\partial u}{\partial t} = \frac{\partial^2 u}{\partial x^2} + \sqrt{u} - u. \qquad (7.3.16)$$

7.3.2 Traveling Wave Solutions

If we denote the TW solutions to Equation (7.3.15) by

$$u(x,t) = g(x - ct)$$

$$= g(z), \quad z = x - ct, \tag{7.3.17}$$

then substituting this in Equation (7.3.16) gives

$$-cg' = g'' + \sqrt{g} - g; \quad g' = \frac{dg}{dz}, \text{etc.}, \tag{7.3.18}$$

which becomes on rearrangement

$$g'' + cg' + \sqrt{g} - g = 0. \tag{7.3.19}$$

Inspection of Equations (7.3.15) and (7.3.19) shows that the introduction of TWs has changed the original problem of studying a nonlinear PDE to one involving an ordinary differential equation. Using the requirements stated in Equations (7.3.9) and (7.3.10), we sketch in Figure 7.5 the two possibilities for the TW profiles. This drawing is based on the fact that the g-ODE has two fixed-points located at

$$\bar{g}^{(1)} = 1, \quad \bar{g}^{(2)} = 0. \tag{7.3.20}$$

The following physical argument will be used to show that TW solutions exist with the general features given in Figure 7.5.

First, relabel the variables

$$z \to t, g \to x, \tag{7.3.21}$$

then Equation (7.3.19) becomes

$$\frac{d^2x}{dt^2} + c\frac{dx}{dt} + \sqrt{x} - x = 0. \tag{7.3.22}$$

This represents in classical mechanics a particle of mass one, acted on by frictional and nonlinear elastic forces, i.e.,

$$\text{Friction force} = -c\frac{dx}{dt}, \tag{7.3.23}$$

$$\begin{pmatrix} \text{Nonlinear} \\ \text{elastic force} \end{pmatrix} = -\sqrt{x} + x. \tag{7.3.24}$$

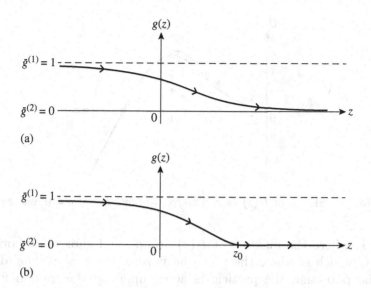

FIGURE 7.5 (a) The TW decreases monotonical from $g(-\infty) = 1$ to $g(+\infty) = 0$, with $g(z)$ becoming zero at $z = \infty$. (b) The TW decreases monotonical from $g(-\infty) = 1$ to $g(z_0) = 0$ and then becomes zero for $z > z_0$.

The nonlinear elastic force is obtainable from the potential energy function

$$U(x) = \left(\frac{2}{3}\right) x^{3/2} - \left(\frac{1}{2}\right) x^2, \tag{7.3.25}$$

$$\left(\begin{array}{c}\text{Nonlinear}\\ \text{elastic force}\end{array}\right) = -\frac{dU(x)}{dx}. \tag{7.3.26}$$

Rewriting Equation (7.3.22) in the system form

$$\frac{dx}{dt} = y, \quad \frac{dy}{dt} = -\sqrt{x} + x - cy, \tag{7.3.27}$$

we observe again that the fixed-points in the $x - y$ plane are located at the positions

$$\left(\bar{x}^{(1)}, \bar{y}^{(1)}\right) = (0, 1), \quad \left(\bar{x}^{(2)}, \bar{y}^{(2)}\right) = (0, 0). \tag{7.3.28}$$

From a sketch of $U(x)$ vs x, see Figure 7.6, it is easily seen that the TW solution corresponds to the particle trajectory, which starts at

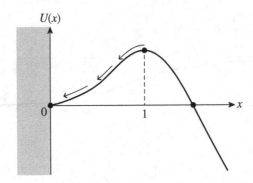

FIGURE 7.6 Sketch of $U(x)$ vs x. The fixed-points are at $x = 0$ and $x = 1$.

$x = 1$, where the maxima of $U(x)$ occurs, and ends at the origin, $x = 0$, which is where the potential is zero. Doing the 'slide' down of the potential, the particle is acted upon by the friction force, $-cy$ ($y = dx/dt$).

Note that this argument plays no restriction on the TW velocity, c, except for $c > 0$.

7.3.3 Traveling Wave Front Behavior

The traveling wave front behavior is determined by what happens when $u(x, t) = g(x - ct) = g(z)$ is very small, i.e.,

$$0 \leq g(z) \ll 1. \qquad (7.3.29)$$

In other words, it is the form of $g(z)$ when $g(z)$ satisfies the conditions given in Equation (7.3.30). This asymptotic behavior can be determined by the use of the method of *dominant balance*, which can be summarized in the following four steps:

(1) Given an equation, first drop all terms that you believe to be 'small' and thus replace the original exact equation with an asymptotic relation.

(2) Second, replace the asymptotic relation with an equation obtained by exchanging the 'asymptotically equal' symbol with the 'equal sign' and solve the resulting equal exactly,

(3) Third, check that the solution obtained in step (2) is fully consistent with the approximations made in Step (1).

(4) If not, use a different set of terms to be kept and others dropped, and start again at step (1).

Applications ■ 181

For our situation, we begin with the full ODE

$$g'' + cg' + \sqrt{g} - g = 0. \quad (7.3.30)$$

Now, we can immediately drop the g term since

$$g \ll \sqrt{g}, \quad 0 < g \ll 1. \quad (7.3.31)$$

Doing this gives

$$g'' + cg' + \sqrt{g} \, 0. \quad (7.3.32)$$

There are three cases to consider

$$\text{Case I:} g'' + cg' = 0, \quad (7.3.33)$$

$$\text{Case II:} g'' + \sqrt{g} = 0, \quad (7.3.34)$$

$$\text{Case III:} cg' + \sqrt{g} = 0. \quad (7.3.35)$$

Let us now assume the following ansatz for the form of $g(z)$ in a neighborhood of the point $z = z_0$,

$$g(z) \sim A(z - z_0)^\alpha, \quad (7.3.36)$$

where (A, α) are unknown parameters to be determined. **Comment** The expression given in Equation (7.3.36) is a short-hand for

$$g(z) \sim \begin{cases} A(z - z_0)^\alpha, & (z_0 - \epsilon) < z \leq z_0, \\ 0, & z > z_0. \end{cases} \quad (7.3.37)$$

Note that the value of z is not essential since Equation (7.3.32) is invariant under the transformation

$$z \to z + z_0. \quad (7.3.38)$$

Under the assumption that we can carry out the indicated differentiations of $g(z)$, as given in Equation (7.3.37), it follows that

$$\begin{cases} g'(z) \sim \alpha A(z - z_0)^{\alpha - 1}, \\ g''(z) \sim \alpha(\alpha - 1) A(z - z_0)^{\alpha - 2}. \end{cases} \quad (7.3.39)$$

We now examine the three cases presented in Equations (7.3.32)–(7.3.34).

7.3.4 Case I

Substitution of the ansatz into Equation (7.3.33) gives

$$\alpha(\alpha - 1) A(z - z_0)^{\alpha-2} + c\alpha A(z - z_0)^{\alpha-1} = 0, \quad (7.3.40)$$

where consistency requires

$$\alpha(\alpha - 1) A + c\alpha A = 0, \quad \alpha - 2 = \alpha - 1. \quad (7.3.41)$$

However, the second relation, involving α, gives a contradiction, $2 = 1$; consequently, Case I is eliminated as determining a valid asymptotic solution for $g(z)$ in the neighborhood of $z = z_0$.

7.3.5 Case II

If Equation (7.3.36) is substituted into Equation (7.3.34), we obtain

$$\alpha(\alpha - 1) A(z - z_0)^{\alpha-2} + A^{\frac{1}{2}}(z - z_0)^{\frac{\alpha}{2}} = 0, \quad (7.3.42)$$

and consistency requires

$$\alpha(\alpha - 1) A + A^{\frac{1}{2}} = 0, \quad \alpha - 2 = \frac{\alpha}{2}. \quad (7.3.43)$$

Solving for α and A gives

$$\alpha = 4, \quad A = \frac{1}{144}. \quad (7.3.44)$$

Comment

Using $\alpha = 4$ in the first expression of Equation (7.3.43), we obtain

$$12A + A^{\frac{1}{2}} = 0, \quad (7.3.45)$$

which can be written

$$12\sqrt{A}\left(\sqrt{A} + \frac{1}{12}\right) = 0. \quad (7.3.46)$$

Selecting the nontrivial solution gives

$$\sqrt{A} = -\left(\frac{1}{12}\right) \to A = \frac{1}{144}. \quad (7.3.47)$$

To check that this $g(z)$, i.e.,

$$g(z) \sim \left(\frac{1}{144}\right)(z - z_0)^4, \quad (7.3.48)$$

is consistent with our level of approximation, we substitute it into Equation (7.3.29) and obtain the expression

$$\left(\frac{1}{12}\right)(z-z_0)^2 + C\left(\frac{1}{36}\right)(z-z_0)^3 - \left(\frac{1}{12}\right)(z-z_0)^2 \qquad (7.3.49)$$

$$-\left(\frac{1}{144}\right)(z-z_0)^4. \qquad (7.3.50)$$

Observe that the first and third terms dominate the asymptotics for $(z-z_0)$ small, and therefore Equation (7.3.48) is consistent with the approximation to Equation (7.3.30) given by Equation (7.3.34). Also, since the first and third terms cancel each other, this expression is asymptotic to zero as $z \to z_0^-$.

7.3.6 Case III

Equation (7.3.35) can be solved exactly. If we impose the condition

$$g(z_0) = 0, \qquad (7.3.51)$$

then this solution is

$$g(z) = \begin{cases} \left(\frac{1}{4}\right)\left[\frac{z_0-z}{c}\right]^2, & z \le z_0; \\ 0, & z > z_0. \end{cases} \qquad (7.3.52)$$

Relative to our assumed 'asymptotic' solution, Equation (7.3.36), it follows that

$$\alpha = 2, A = \frac{1}{4c^2}, \qquad (7.3.53)$$

Substituting Equation (7.3.52) into Equation (7.3.30) gives the expression

$$\left(\frac{1}{2c^2}\right)(z-z_0)^0 + \left(\frac{1}{2c}\right)(z-z_0)$$

$$-\left(\frac{1}{2c}\right)(z-z_0) - \left(\frac{1}{4c^2}\right)(z-z_0)^2, \qquad (7.3.54)$$

where in the third term we used

$$\sqrt{A} = -\left(\frac{1}{2c}\right). \qquad (7.3.55)$$

Note that the first term dominates the three other terms, thus this asymptotic solution is not consistent.

In summary, only the Case II asymptotic solution is consistent with the behavior of $g(z)$ in a neighborhood of $z = z_0$.

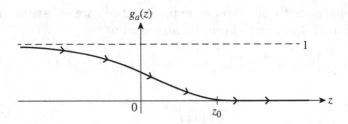

FIGURE 7.7 Sketch of $g_a(z)$ vs z for the function given in Equation (7.3.58).

7.3.7 Approximation to Traveling Wave Solution

The Heaviside step 'function' is defined to be

$$H(x) = \begin{cases} 1, & x \geq 0, \\ 0, & x < 0. \end{cases} \qquad (7.3.56)$$

Using this function, then an *approximate solution* for the *traveling wave problem* defined by Equations (7.3.5) and (7.3.8), and satisfying the conditions in Equations (7.3.9) and (7.3.10), is

$$u(x,t) = g(x - ct) \simeq g_a(z), \qquad z = x - ct, \qquad (7.3.57)$$

where

$$g_a(z) = \left\{ 1 - \exp\left[-\frac{(z - z_0)^4}{144} \cdot H(z_0 - z) \right] \right\}. \qquad (7.3.58)$$

As stated previously, the value of this approximate solution for practical applications will depend on the particular needs of the user and what they hope to achieve.

Figure 7.7 is a sketch of the curve representing $g_a(z)$.

Finally, we give a summary of the general features of the TW solution of Equation (7.3.15), which also holds for its approximation, $g_a(z)$:

(i) For arbitrarily large, negative values of z, the TW solution has a value that starts at $g(-\infty) = 1$ and then monotonically decreases.

(ii) At some $z = z_0$, $g(z_0) = 0$ and $g(z) = 0$ for $z > z_0$.

(iii) At $z = z_0$, the slope is zero, i.e.,

$$\frac{dg(z_0)}{dz} = 0, \qquad (7.3.59)$$

with

$$\frac{dg(z)}{dz} = 0, \quad z > z_0. \qquad (7.3.60)$$

(iv) The result (iii) implies that $g(z)$ is a piece-wise-continuous function that is smooth over the interval, $-\infty < z < +\infty$.

7.4 COMMENTS ON FUNCTIONAL EQUATION MODELS OF RADIOACTIVE DECAY AND HEAT CONDUCTION

We now discuss briefly the selection of which mathematical structures to use for the modeling of a particular physical phenomenon.

- Radioactive decay
- Heat conduction (in 1-dim)

When we say 'mathematical structure', we mean the type of mathematics used to construct a model. Examples of such structures are

- Differential equations
- Difference equations
- Integral equations
- Algebraic equations
- Combinations of such equations etc.

Most mathematical models (MM) are formulated in terms of differential equations. A major reason is that differential equations are but the extension of calculus, a subject all of us as scientists and mathematicians have taken in our training at the university. Few of us are familiar with difference equations, and consequently, MMs involving these structures are lesser in number.

Another essential point is that in most scientific modeling activities, the assumption is implicitly made that space–time is 'continuous'. Note that this is, in some sense, an extreme assumption since the maneuver in which we actually probe physical systems is always discrete in both space and time, i.e., physical measurements are always done discretely. However, this assumption allows us to immediately make use of differential equations.

There is also the issue with the use of discrete models based on difference equations. In general, they are just much harder to construct. And, few scientists fully understand that difference equations are not approximations to differential equations; they have their own dynamics that can differ remarkably from differential equations.

For example, consider the linear, first-order ODE

$$\frac{dx}{dt} = -\lambda x, \quad x(0) = x_0, \quad \lambda > 0, \tag{7.4.1}$$

which provides an MM for many physical phenomena. Its solution is

$$x(t) = x_0 e^{-\lambda t}. \tag{7.4.2}$$

Let us now discrete the ODE using a simple finite-difference 'approximation', i.e.,

$$t \to t_k = hk, \quad k = (0, 1, 2, \ldots); \quad h = \Delta t; \tag{7.4.3}$$

$$x(t) \to x(t_k) \to x_k, \tag{7.4.4}$$

$$\frac{dx(t)}{dt} \to \frac{x_{k+1} - x_k}{h}, \tag{7.4.5}$$

which gives

$$\frac{x_{k+1} - x_k}{h} = -\lambda x_k, \quad x_0 \text{ given}; \tag{7.4.6}$$

or

$$x_{k+1} = (1 - \lambda h) x_k. \tag{7.4.7}$$

The last equation has the solution

$$x_k = x_0 (1 - \lambda h)^k. \tag{7.4.8}$$

Note that all the solutions to the ODE smoothly decrease to zero. However, the solutions to the discretization, Equation (7.4.8), have this behavior only if

$$0 < \lambda h < 1. \tag{7.4.9}$$

Otherwise, the solutions, x_k, oscillate in values and can become unbounded for sufficiently large h. Thus, h can be taken as a bifurcation parameter with its critical value at

$$h_c = \frac{1}{\lambda}. \qquad (7.4.10)$$

Also, it is easily seen that, in general,

$$x_k \neq x(t_k), \qquad (7.4.11)$$

where $x(t)$ is the solution to the differential equation and x_k is the solution to the discrete equation.

Now, let us try something that will turn out to be very interesting. Start with the exact solution to the ODE, Equation (7.4.2), and discretize it to obtain

$$x(t) = x_0 e^{-\lambda t} \longrightarrow x_k = x_0 e^{-\lambda t_k}. \qquad (7.4.12)$$

It follows that

$$x_{k+1} = \left(x_0 e^{-\lambda t_k}\right) e^{-\lambda h} \qquad (7.4.13)$$

or

$$x_{k+1} = \left(e^{-\lambda h}\right) x_k. \qquad (7.4.14)$$

But

$$x_{k+1} - x_k = -\left(1 - e^{-\lambda h}\right) x_k \qquad (7.4.15)$$

and

$$x_{k+1} - x_k = -\left(\frac{1 - e^{-\lambda h}}{\lambda}\right)(\lambda x_k), \qquad (7.4.16)$$

and, finally,

$$\frac{x_{k+1} - x_k}{\left(\frac{1 - e^{-\lambda h}}{\lambda}\right)} = -\lambda x_k. \qquad (7.4.17)$$

Following how this last equation was constructed, we can conclude that it is an *exact finite-difference discretization* of the ODE given in Equation (7.4.1), i.e., exact in the sense that

$$x_k = x(t_k). \qquad (7.4.18)$$

188 ■ Introduction to Qualitative Methods for Differential Equations

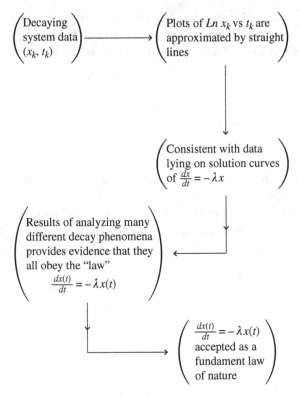

FIGURE 7.8 The discovery path for the formulation of the exponential decay law, $dx/dt = -\lambda x$, and its acceptance as fundamental.

Since scientists and mathematicians of ten discretize ODEs for the purposes of determining numerical solutions and since 'numerical instabilities' may occur, we can conclude that these numerical instabilities exist because improper discretization was used. The difficulty is that, in general, we do not know how to construct 'exact finite-difference schemes' or even if such schemes exist for any given ODE or PDE. Figure 7.8 gives a possible path to the assessment that $dx/dt = -\lambda x$ is a 'fundamental' law of nature for decay phenomena.

But, another path is also plausible and could have been taken. This alternative path is based on the concept of half-life. Experimental evidence shows that for direct decaying systems, there exists a time, τ, such that if at time t the decaying substance has 'activity' $x(t)$, then at time, $t + \tau$, the activity is one-half of its value at time

t. Mathematically, these results can be summarized by the equation

$$x(t+\tau) = \left(\frac{1}{2}\right)x(t). \tag{7.4.19}$$

This equation is not a differential equation; it is called a functional equation. It turns out that, in general, *functional equations* are much more difficult to solve than differential equations. Further, these equations generally do not have general solutions for which every solution is a particular case.

The paper by Mickens and Rucker (2023) calculates the following solution to this particular functional equation that is relevant to decay processes; it is

$$x(t) = A(t)\exp\cdot\left[-\left(\frac{\text{Ln}(2)}{\tau}\right)t\right], \tag{7.4.20}$$

where $A(t)$ is an arbitrary periodic function of period, τ, i.e.,

$$A(t+\tau) = A(t), \tag{7.4.21}$$

and it is odd, i.e.,

$$A(-t) = -A(t). \tag{7.4.22}$$

If we write $A(t)$ as

$$A(t) = x_0 e^{\theta(t)}, \tag{7.4.23}$$

then $\theta(t)$ can be expressed as a sine-Fourier series,

$$\theta(t) = \sum_{k=1}^{\infty} b_k \sin\left[\left(\frac{2\pi k}{\tau}\right)t\right]. \tag{7.4.24}$$

If only one term is retained in the Fourier series, then

$$x(t) = x_0 \exp\left\{b_i \sin\left[\left(\frac{2\pi}{\tau}\right)t\right]\right\} \cdot \exp\left[-\left(\frac{Ln(2)}{\tau}\right)t\right] \tag{7.4.25}$$

and taking the logarithm of this gives

$$\text{Ln}(X(t)) = \text{Ln}(x_0) + b_1 \sin\left[\left(\frac{2\pi}{\tau}\right)t\right] - \left[\frac{\text{Ln}(2)}{\tau}\right]t. \tag{7.4.26}$$

Note that a plot of $\text{Ln}(x(t))$ vs t produces a straight line with oscillations. This oscillatory line will have a slope of value

$$\text{Slope} = \frac{\text{Ln}(2)}{\tau}, \tag{7.4.27}$$

and the oscillations will only be observable at small t since the third term dominates for large values of t. This behavior is consistent with some experiments. See the references in this section.

The main point of the above discussion is that radioactive decay can be modeled in more than one way and that a change in the mathematical structure here from differential equations to functional equations provides a possible deeper understanding of the physical phenomena being investigated. For this case, decaying systems, we expect a future quantum-based calculation to resolve the various issues raised by various experiments.

We turn now to simple heat conduction in one-space dimension. At the 'macroscopic level', the governing PDE is

$$u_t = Du_{xx}, \quad u = u(x,t), \quad D = \text{constant} > 0. \qquad (7.4.28)$$

The corresponding 'microscopic level' equation is that for a random walk

$$u(x, t+\tau) = \left(\frac{1}{2}\right) u(x+a, t) + \left(\frac{1}{2}\right) u(x-a, t), \qquad (7.4.29)$$

where (τ, a) are fixed time and space parameters. To see the link between Equations (7.4.28) and (7.4.29), we rewrite the random walk equation in the following ways:

$$\begin{cases} u(x, t+\tau) = u(x,t) = \left(\frac{1}{2}\right)[u(x+a,t) - 2u(x,t) + u(x-a,t)], \\ \dfrac{u(x,t+\tau) - u(x,t)}{\tau} = \left(\dfrac{a^2}{2\tau}\right) \dfrac{[u(x+a,t) - 2u(x,t) + u(x-a,t)]}{a^2} \end{cases}$$
$$(7.4.30)$$

We now make a 'huge' assumption. In place of the fixed parameters; (τ, a), we assume that it makes 'physical sense' to take their magnitudes to zero, i.e.,

$$\tau \to 0, \quad a \to 0, \qquad (7.4.31)$$

but this is to be done such that

$$\frac{a^2}{2\tau} = D \text{ is constant} > 0. \qquad (7.4.32)$$

If we do this, then the calculus informs us that partial derivatives will arise from our manipulations, i.e.,

$$\lim_{\tau \to 0} \frac{u(x, t+\tau) - u(x,t)}{\tau} = \frac{\partial u(x,t)}{\partial t}, \qquad (7.4.33)$$

$$\underset{a\to 0}{\text{Lim}}\frac{u(x+a,t)-2u(x,t)+u(x-a,t)}{a^2} = \frac{\partial^2 u(x,t)}{\partial x^2}. \qquad (7.4.34)$$

With these items in mind, we then determine that Equation (7.4.30) becomes

$$\frac{\partial u(x,t)}{\partial t} = D\frac{\partial^2 u(x,t)}{\partial x^2}, \qquad (7.4.35)$$

which is the heat equation given in Equation (7.4.28).

Again, to derive the heat PDE from the random walk equation, we make another 'huge' assumption, namely that $u(x,t)$ has Taylor series at each point, (x,t), of its domain of definition. While this may or not hold for the actual physical universe, we need this assumption for the derivation.

Also, the parameters, (τ, a), play very different roles in the random walk and heat equations. For the random walk equation, these two parameters are in principle independent of each other. We may consider them to be the 2-dim lattice parameters in a 2-dim discrete space–time. However, in the heat equation PDE, these parameters do not explicitly appear and their values are zero, i.e., $a = 0$ and $\tau = 0$. This is because

$$a \to 0, \tau \to 0, \frac{a^2}{2\tau} = D = \text{constant}. \qquad (7.4.36)$$

This fact should cause us to pause and ask: Why should we believe that these kinds of arguments are valid? However, it must be recognized that scientists for the previous century and a half have been doing just this and, as a consequence, they created the modern technological world. We have been able to construct mathematics-based approximate solutions to physical problems without the need to understand the whole underlying mathematical theory.

A major goal of physical theory is to construct a mathematical structure that incorporates within itself the constraints of 'real physics', such as the discrete nature of doing physical measurements.

Finally, we will complete this section by examining the random walk equation (which we write again)

$$u(x, t+\tau) = \left(\frac{1}{2}\right)u(x+a,t) + \left(\frac{1}{2}\right)u(x-a,t). \qquad (7.4.37)$$

A minor generalization of this equation is

$$u(x,t+\tau) = pu(x+a,t) + (1-2p)u(x,t) + pu(x-a,t), \qquad (7.4.38)$$

where
$$0 < p \le \frac{1}{2}. \tag{7.4.39}$$

This equation can be rewritten as
$$\frac{u(x, t+\tau) - u(x, t)}{\tau} = \left(\frac{pa^2}{\tau}\right)\left[\frac{u(x+a, t) - 2u(x, t) + u(x-a, t)}{a^2}\right], \tag{7.4.40}$$

and gives in the limits
$$a \to 0, \tau \to 0, \frac{pa^2}{\tau} = D = \text{constant} > 0, \tag{7.4.41}$$

again the heat conduction PDE,
$$\frac{\partial u(x,t)}{\partial t} = D\frac{\partial^2 u(x,t)}{\partial x^2}. \tag{7.4.42}$$

The separation-of-variables solutions to the heat PDE are well known, i.e.,
$$u(x, t) = F(x)\, G(t) \tag{7.4.43}$$

and
$$\frac{G'(t)}{DG(t)} = \frac{F''(x)}{F(x)} = -k^2, \tag{7.4.44}$$

where the separation constant is taken to be $(-k^2)$.
The solutions to the ODEs in Equation (7.4.44)
$$G'(t) = -(k^2 D)\, G(t), \tag{7.4.45}$$

$$F''(x) = -k^2 F(x), \tag{7.4.46}$$

are
$$G(t) = A(k^2)\, e^{-(k^2 D)t}, \tag{7.4.47}$$

$$F(x) = B_1(k^2)\sin(kx) + B_2(k^2)\cos(kx), \tag{7.4.48}$$

and therefore,
$$u(x, t, k^2) = F(x) G(t) = [A_1(k^2) \sin(kx) + A_2(K^2) \cos(kx)], \quad (7.4.49)$$

where
$$A_1(k^2) = A(k^2) B_1(k^2), \quad A_2(k^2) = A(k^2) B_2(k^2). \quad (7.4.50)$$

The 'general solution' based on the separation-of-variables method is therefore

$$u(x, t) = \int u(x, t, k^2) \quad (7.4.51)$$

or, in detail,

$$u(x, t) = \int e^{-(k^2 D)t} [A_1(k^2) \sin(kx) + A_2(k^2) \cos(kx)] \, dk. \quad (7.4.52)$$

Let us now see if we can do the same SOV procedure for the randow walk equation

$$u(x, t+\tau) = \left(\frac{1}{2}\right) u(x+a, t) + \left(\frac{1}{2}\right) u(x-a, t), \quad (7.4.53)$$

with
$$u(x, t) = F(x) G(t). \quad (7.4.54)$$

Substituting this assumed SOV solution into Equation (7.4.53) gives

$$F(x) G(t+\tau) = \left(\frac{1}{2}\right) F(x+a) G(t) + \left(\frac{1}{2}\right) F(x-a) G(t), \quad (7.4.55)$$

and on separating the variables the expression

$$\frac{G(t+\tau)}{G(t)} = \frac{F(x+a) + F(x-a)}{2 F(x)}. \quad (7.4.56)$$

Since we want to have $G(t)$ to be a decreasing function of t, the separation constant will be selected as $\exp[-\lambda^2]$, $-\infty < \lambda < +\infty$. Therefore, the time and space equations are

$$G(t+\tau) = e^{-\lambda^2} G(t), \quad (7.4.57)$$

$$F(x+a) - 2\left(e^{-\lambda^2}\right) F(x) + F(x-a) = 0. \quad (7.4.58)$$

The first equation has the solution

$$G(t) = \theta(t) \exp\left[-\lambda^2 \left(\frac{t}{\tau}\right)\right], \qquad (7.4.59)$$

where $\theta(t)$ is an arbitrarily bounded, periodic function, i.e.,

$$\theta(t+\tau) = \theta(t) \qquad (7,4.58)$$

Likewise, the 'general' solution for $F(x)$ is the following expression (which I derived from the listing in Polyanin (1998)) book,

$$F(x) = \theta_1(x) \cos\left[\frac{\varphi(\lambda^2) x}{a}\right] + \theta_2(x) \sin\left[\frac{\varphi(\lambda^2) x}{a}\right], \qquad (7.4.60)$$

where $\theta_1(x)$ and $\theta_2(x)$ are bounded, periodic functions, i.e.,

$$\theta_i(x+a) = \theta_i(x), \quad i = (1,2), \qquad (7.4.61)$$

with $\varphi(\lambda^2)$ defined as

$$\tan \varphi(\lambda^2) = \left[e^{\lambda^2} - 1\right]^{\frac{1}{2}}. \qquad (7.4.62)$$

Putting all these together, we see that the expression for the SOV solution is relatively complex, i.e.,

$$U(x, t, \lambda^2) = [\text{Equation}\,(7.4.57)] \cdot [\text{Equation}\,(7.4.59)]. \qquad (7.4.63)$$

Finally, the full SOV solution to the random walk equation, see Equation (7.4.53), is

$$u(x,t) = \int u(x, t, \lambda^2)\, d\lambda^2, \qquad (7.4.64)$$

and, at this point, we stop.

7.5 APPROXIMATE SOLUTIONS TO A MODIFIED, NONLINEAR MAXWELL–CATTANE EQUATION

Experimental and theoretical work on heat conduction shows clearly that the simple heat equation

$$u_t(x,t) = D u_{xx}, \quad u = u(x,t), D > 0, \qquad (7.5.1)$$

does not give results in agreement with the data. A major difficulty with this equation is that it makes the prediction that the speed of information transfer is infinite. A way to resolve this particular issue is to add an additional term, τu_{tt}, to give the PDE

$$\tau u_{tt} + u_t = D u_{xx}, \qquad (7.5.2)$$

where the positive parameter, τ, is usually interpreted as some type of relaxation time in the thermal system. It should be noted that Equation (7.5.2) is a damped, linear wave equation with the speed of signal or information transfer given by

$$c = \sqrt{\frac{D}{\tau}}. \qquad (7.5.3)$$

The goals of this section are to investigate the solutions to a modified version of Equation (7.5.2), where τ is now a function of u. The function selected is

$$\tau(u) = \frac{1}{a\sqrt{u}}, \qquad (7.5.4)$$

where a is a positive parameter. This particular form for $\tau(u)$ is suggested by the ideal gas law, i.e., if u is the Kelvin temperature, then for fixed volume

$$\langle \text{Kinetic energy}^2 \rangle \propto u, \qquad (7.5.5)$$

where the symbol, $\langle \cdots \rangle$ denotes an average. But, if all the molecules have the same mass, the

$$\langle \text{Kinetic energy} \rangle \propto \langle \text{Velocity} \rangle. \qquad (7.5.6)$$

Using physical units, it follows that

$$\left\langle \left(\frac{\text{Length}}{\text{Time}}\right)^2 \right\rangle \propto u. \qquad (7.5.7)$$

Therefore, if we identify the 'time scale' with the parameter, τ, then the result in Equation (7.5.4) is found. We now assume that such a relationship holds for our system of interest, regardless of its state as a gas, liquid or solid.

With this $\tau(u)$, Equation (7.5.2) becomes

$$\left(\frac{1}{a\sqrt{u}}\right) u_{tt} + u_t = D u_{xx}, \qquad (7.5.8)$$

and this can be rewritten to get the expression

$$u_{tt} + a\sqrt{u}\,(u_t - Du_{xx}) = 0. \tag{7.5.9}$$

The remainder of this section will be devoted to a study of various exact and approximate solutions to Equation (7.5.9). Please remember that we are not claiming that Equation (7.5.9) is applicable to any actual physical system. We are studying it, for now, as an interesting generalization of the damped wave equation.

Finally, it should be stated that in most of the research literature, Equation (7.5.2) is called the Maxwell–Cattaneo equation. Thus, Equation (7.5.9) can be thought of as a nonlinear Maxwell–Cattaneo PDE.

7.5.1 Positivity and Equilibrium Solutions

In physics, the temperature of a system always refers to its value in the Kelvin absolute scale. With this scale, the lowest temperature is zero and all physical realizable temperatures have positive value. So if u is identified with the temperature, then

$$u(x,t) \geq 0. \tag{7.5.10}$$

Inspection of our modified, nonlinear, Maxwell–Cattaneo equation (MNMCE) also indicates that unless this condition holds, the solutions will be complex valued.

Since each term in the MNMCE contains a derivative, then

$$u(x,t) = \bar{u} > 0, \tag{7.5.11}$$

where $\bar{u}$ is a constant in a solution. However, it must be stated that classical thermodynamics is formulated such that absolute zero Kelvin can only be attained asymptotically, i.e., to reach zero temperature requires an infinite iterative physical process.

In general, equilibrium solutions are those for which

$$\frac{\partial}{\partial t} u(x,t) = 0. \tag{7.5.12}$$

Therefore,

$$u_{eq}(x,t) = A + Bx, \tag{7.5.13}$$

where we have used the result in Equation (7.5.12) to derive the relation:

$$Du_{xx}(x, t) = 0, \qquad (7.5.14)$$

from Equation (7.5.9). The constants A and B are arbitrary and depend on the boundary conditions of each particular problem; but, they must always be selected such that

$$u_{eg}(x, t) \geq 0. \qquad (7.5.15)$$

Note that this is also a solution to the usual linear heat equation, $u_t = Du_{xx}$. In particular, consider a thin, insulated wire of length, L, with one end held at $u = T_0 > 0$, at $x = 0$, and the other end fixed at $x = L$, at temperature $u = T_L > 0$. The equilibrium solution is

$$u_{eq}(x, t) = T_0 + \left(\frac{T_L - T_0}{L}\right)x, \qquad 0 \leq x \leq L. \qquad (7.5.16)$$

In summary, the *equilibrium solutions* are the time-independent solutions; they can depend on x, but only linearly.

7.5.2 Space-Independent Solutions

The space-independent solutions

$$u(x, t) = S(t); \quad S(t) > 0, t \geq 0, \qquad (7.5.17)$$

satisfy the following ODE

$$S'' + a\sqrt{S}S' = 0, \quad S' = \frac{dS}{dt}. \qquad (7.5.18)$$

(We remind ourselfies that $a > 0$.) Integrating once gives

$$S' + \left(\frac{2a}{3}\right)S^{\frac{3}{2}} = A. \qquad (7.5.19)$$

There are three cases to examine,

$$A < 0, \quad A = 0, \quad A > 0. \qquad (7.5.20)$$

Case (i): $A<0$

For this case, write A as

$$A = -|A|. \qquad (7.5.21)$$

To solve Equation (7.5.19), we must evaluate an integral of the form

$$I(w) = \int \frac{dw}{1+w^{3/2}} = (?), \qquad (7.5.22)$$

and this turns out to have the following evaluation, which cannot be solved explicitly for w,

$$I(w) = \left(\frac{1}{3}\right)\left[\operatorname{Ln}\left(x - \sqrt{x} + 1\right) - 2\operatorname{Ln}\left(\sqrt{x} + 1\right)\right]$$
$$+ \left(\frac{2}{\sqrt{3}}\right)\arctan\left(\frac{2\sqrt{x}-1}{\sqrt{3}}\right). \qquad (7.5.23)$$

Clearly, we need to use another approach to determine meaningful information on the main features of the solutions to Equation (7.5.19) when $A < 0$.

We proceed by looking at Equation (7.5.19) in more detail. From the perspective of the physics of a system modeled by this first-order, nonlinear, ODE, the following conditions are expected to hold,

$$\begin{cases} \cdot \operatorname{Lim}_{t\to\infty} S(t) = S^* > 0. \\ \cdot \operatorname{Lim}_{t\to\infty} S'(t) = 0. \end{cases} \qquad (7.5.24)$$

As a result of these conditions, the constant A is given by the expression

$$\left(\frac{2a}{3}\right)(S^*)^{\frac{3}{2}} = A > 0. \qquad (7.5.25)$$

This means that Case (i), $A < 0$, and Case (ii), $A = 0$, do not need to be considered. With this in mind, we can rewrite Equation (7.5.19) to the form

$$S' = A - \left(\frac{2a}{3}\right)S^{3/2}$$

$$= \left(\frac{2a}{3}\right)\left[(S^*)^{3/2} - S^{3/2}\right]. \qquad (7.5.26)$$

Also, it must not be forgotten that

$$S(t) = 0, \qquad (7.5.27)$$

is a nontrivial solution to Equation (7.5.18). In the $S(t)$ vs t, the solution plane, Figure 7.9 sketches the types of solution behaviors.

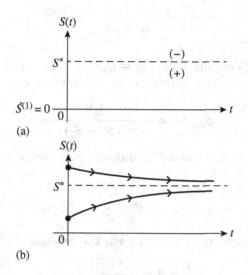

FIGURE 7.9 Sketch of solution plane for Equation (7.5.18).

(a) If $0 < S_0 < S^*$, then $S(t)$ increases smoothly to s^*.
(b) If $S_0 > S^*$, then $S(t)$ decreases smoothly to s^*.
(c) If $S_0 = S^*$, then $S(t)$ stays at this value.

Clearly, the two fixed-points have the following stability properties

$$\begin{cases} \bar{S}^{(1)} = 0: \text{unstable}, \\ \bar{S}^2 = S^*: \text{stable}. \end{cases} \quad (7.5.28)$$

An (almost) ad hoc form that can be used to approximate solutions to Equation (7.5.26) is

$$S_{\text{app}}(t) = \frac{A}{1 + Be^{-\lambda t}}. \quad (7.5.29)$$

This particular structure is hinted at by looking at the so-called logistic equation

$$\frac{dy}{dt} = \lambda_1 y - \lambda_2 y^2, \quad \lambda_1 > 0, \lambda_2 > 0. \quad (7.5.30)$$

Its exact solution is

$$y(t) = \frac{y_0 y^*}{y_0 + (y^* - y_0) e^{-\lambda_1 t}}, \quad (7.5.31)$$

$$y^* = \frac{\lambda_1}{\lambda_2}, \tag{7.5.32}$$

where λ_1 and λ_2 are given and $y_0 = y(0) > 0$. For our ODE, Equation (7.5.26), we use as our approximate solution the expression

$$S_{\text{app}}(t) = \frac{S_0 S^*}{S_0 + (S^* - S_0) e^{-\lambda t}}, \tag{7.5.33}$$

where λ is to be determined. Using the fact that λ is related to the time scale,

$$t^* = \frac{1}{\lambda}, \tag{7.5.34}$$

then this allows us to calculate λ via the relation

$$\frac{1}{t^*} = \lambda \equiv \left| \frac{dF(S)}{dS} \right|_{S = S^*}, \tag{7.5.35}$$

where

$$F(S) = \left(\frac{2a}{3}\right)\left[(S^*)^{\frac{3}{2}} - S^{3/2}\right]. \tag{7.5.36}$$

(A discussion of the basis of this procedure is given in Mickens (2022), Section 0.5.) Therefore,

$$\lambda = a\sqrt{S^*}. \tag{7.5.37}$$

In the article by Herron and Mickens (2023), they found that $S_{\text{app}}(t)$ was remarkably close to an accurate numerical solution of Equation (7.5.26).

7.5.3 Traveling Waves

We now investigate whether the MNMCE

$$u_{tt} + a\sqrt{u}\,(u_t - Du_{xx}) = 0, \tag{7.5.38}$$

has traveling wave (TW) solutions, i.e.,

$$u(x, t) = f(z), \quad z = x - ct. \tag{7.5.39}$$

Substituting this expression into Equation (7.5.38) gives

$$c^2 f'' + a\sqrt{f}\,(-cf' - Df'') = 0, \tag{7.5.40}$$

or
$$(c^2 - aD\sqrt{f})f' = (ac)\sqrt{f}f. \qquad (7.5.41)$$

Note that we are only interested in solutions with the property
$$f(z) > 0, \quad -\infty < z < +\infty. \qquad (7.5.42)$$

From Equation (7.5.42), we can construct its first-order system representation,
$$\frac{df}{dz} = f, \quad \frac{df'}{dz} = \frac{(ac)\sqrt{f}f'}{c^2 - (aD)\sqrt{f}}. \qquad (7.5.43)$$

For the remainder of this section, we assume
$$C > 0. \qquad (7.5.44)$$

The case $c < 0$ can be worked out in a manner similar to the case $c > 0$.

The trajectories in the f-f' phare-plane are solutions to the following first-order, nonlinear ODE
$$\frac{df'}{df} = \frac{(ac)\sqrt{f}}{c^2 - (aD)\sqrt{f}}. \qquad (7.5.45)$$

The corresponding nullclines are
$$\frac{df'}{df} = 0 : \begin{cases} (1) \text{ along the } f'\text{-axis,} \\ (2) \text{ along the } f\text{-axis,} \end{cases} \qquad (7.5.46)$$

$$\frac{df'}{df} = \infty : \begin{cases} (1) \text{ along the } f\text{-axis,} \\ (2) \text{ along } f = \left(\frac{c^2}{aD}\right)^2. \end{cases} \qquad (7.5.47)$$

Figure 7.10 provides in (a) information on the properties of the f–f' phase-space. Some of these features are:

(i) The nullclines divide the phase-space into four regions (for $f > 0$). They are indicated by the notation (R_1, R_2, R_3, R_4).

(ii) The arrows coming out of a representative point for each region indicate the directions of the trajectories in each region. In more detail, we have

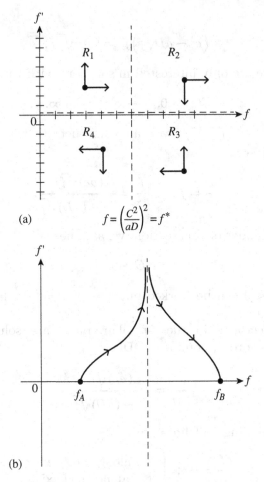

FIGURE 7.10 (a) Nullclines for Equations (7.5.43) with $c > 0$. Horizontal-dashed lines (—) and vertical-dashed lines (), indicate, respectively, where the slopes of the trajectories are zero and infinite. (b) A traveling wave trajectory.

$$R_1 \; : \; 0 < f < f^*, f' > 0, (\nearrow);$$

$$R_2 \; : \; f > f^*, f' > 0, (\searrow);$$

$$R_3 \; : \; f > f^*, f' < 0, (\searrow);$$

$$R_4 \; : \; 0 < f < f^*, f' < 0, (\nearrow).$$

(iii) The positive f-axis is both a zero- and infinite nullcline. Therefore, every point on the positive f-axis is a fixed-point, i.e., constant solution.

(iv) The fixed-points on the f-axis interval

$$0 < f < f^*, \qquad (7.5.48)$$

are all unstable, while the fixed-points on the f-axis interval

$$f > f^*, \qquad (7.5.49)$$

are stable.

(v) The (b) part of Figure 7.10 gives a sketch of a typical traveling wave solution. It starts at f_A, i.e.,

$$f_A = f(-\infty) > 0 \qquad (7.5.50)$$

and increases to f_B, i.e.,

$$f_B = f(+\infty) > 0, \qquad (7.5.51)$$

$$f_A < f_B. \qquad (7.5.52)$$

At a value of z call it $z = z^*$, where

$$\frac{df}{df} = \infty \qquad (7.5.53)$$

the slope in $f(z)$ has a 'kink'. Figure 7.11 sketches a typical traveling wave solution.

(vi) For the situation where $c < 0$, then all of the trajectory directional flow arrows have their f' signs flip and the general $f\text{-}f$ phase-space and traveling waves take the forms sketched in Figure 7.11.

Now the traveling waves start at f_c, i.e.,

204 ■ Introduction to Qualitative Methods for Differential Equations

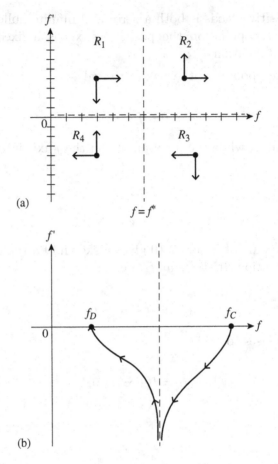

FIGURE 7.11 (a) Nullclines for Equation (7,5.43), with $c < 0$. Notations are the same as in Figure 7.10.

$$f_c = f(-\infty) > 0, \tag{7.5.54}$$

and decreases to f_D; i.e.,

$$f_D = f(+\infty) > 0, \tag{7.5.55}$$

$$f_D < f_c. \tag{7.5.56}$$

(vii) The traveling wave solutions are drawn in Figure 7.12.

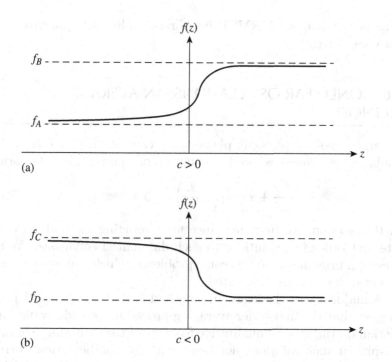

FIGURE 7.12 Representations of the traveling wave solutions for Equation (7.5.43).

7.5.4 Resume

We have derived a number of the important aspects of a nonlinear generalization of the Maxwell–Cattaneo PDE. In particular, the following results were obtained:

(a) Non-negative solutions were shown to exist.

(b) Explicit forms were calculated for the equilibrium solutions.

(c) For the space-independent solutions, we were not able to solve explicitly their functional behavior, but we derived their significant properties and constructed an approximation to these solutions, which was 'accurate' when compared to corresponding numerical solutions.

(d) The investigated traveling wave solutions, mainly using 2-dim phase-space techniques, showed that they exist. However, at the level of our calculations, the value of their speed, c, could not be calculated.

Whether or not our MNMCE has direct a physical application was not considered.

7.6 NONLINEAR OSCILLATIONS: AN AVERAGING METHOD

A large class of physical phenomena can be modeled as 1-dim, nonlinear oscillators where the equation of motion takes the form

$$\frac{d^2x}{dt^2} + x = \epsilon f\left(x, \frac{dx}{dt}\right), \quad 0 < \epsilon \ll 1. \tag{7.6.1}$$

In this section, we first introduce an 'averaging method' to calculate first order in ϵ solutions to such differential equations. We also present a brief discussion on some problems, which may arise if terms higher-order in ϵ are calculated.

A major use of the averaging method is that there is evidence to suggest that the first-order averaging procedure provides valid information on the general qualitative features of the oscillatory solutions in spite of such solutions not being entirely 'mathematical correct'. Consequently, the system can be examined and understood without having a rigorously derived exact solution.

Section 7.6.1 gives a heuristic derivation of the method of averaging, suitable for most physical scientists. This is followed by Section 7.6.2 which discusses some limitations of the averaging method as indicated by a system with limit-cycles. Finally, in Section 7.6.3, we work out the details of applying the method of averaging to several explicit examples.

7.6.1 First Approximation of Krylov and Bogoliubov

To begin, let $\epsilon = 0$. This reduces Equation (7.6.1) to the form

$$\frac{d^2x}{dt^2} + x = 0 \tag{7.6.2}$$

whose general solution is

$$x(t) = a \cos(t + \varphi), \tag{7.6.3}$$

with

$$\frac{dx(t)}{dt} = -a \sin(t + \varphi), \tag{7.6.4}$$

where (a, φ) are arbitrary constants.

For $0 < \epsilon \ll 1$, assume that Equation (7.6.1) has a solution that can be represented as

$$x(t) = a(t) \cos[t + \varphi(t)], \qquad (7.6.5)$$

where a and φ are now functions of t and ϵ, i.e.,

$$a \to a(t, \epsilon) = a(t), \qquad (7.6.6)$$

$$\varphi \to \varphi(t, \epsilon) = \varphi(t). \qquad (7.6.7)$$

Let us also assume that the derivative of $x(t)$ has the form

$$\frac{dx(t)}{dt} = -a(t) \sin[t + \varphi(t)]. \qquad (7.6.8)$$

Note that we have the freedom to do this since $a(t)$ and $\varphi(t)$ are the present unknown.

If we take the derivative of Equation (7.6.5)

$$\frac{dx}{dt} = \frac{da}{dt} \cos \psi - a \sin \psi - a \frac{d\varphi}{dt} \sin \psi, \qquad (7.6.9)$$

where

$$\psi(t) = t + \varphi(t), \qquad (7.6.10)$$

and require

$$\frac{da}{dt} \cos \psi - a \frac{d\varphi}{dt} \sin \psi = 0, \qquad (7.6.11)$$

then the relation in Equation (7.6.8) is obtained, along with a relationship between the derivatives da/dt and $d\varphi/dt$, given in Equation (7.6.11).

If Equation (7.6.8) has its derivative taken, we have

$$\frac{d^2x}{dt^2} = -\frac{da}{dt} \sin \psi - a \cos \psi - a \frac{d\varphi}{dt} \cos \psi. \qquad (7.6.12)$$

Substituting Equations (7.6.5), (7.6.8) and (7.6.12) into Equation (7.6.1) gives

$$\frac{da}{dt} \sin \psi + a \frac{d\varphi}{dt} \cos \psi = -\epsilon f(a \cos \psi, -a \sin \psi), \qquad (7.6.13)$$

which is linear in the derivatives, da/dt and $d\varphi/dt$. Solving Equations (7.6.11) and (7.6.13) for these derivatives gives the results

$$\frac{da}{dt} = -\epsilon f(a\cos\psi, -a\sin\psi)\sin\psi, \qquad (7.6.14)$$

$$\frac{d\varphi}{dt} = -\left(\frac{\epsilon}{a}\right) f(a\cos\psi, -a\sin\psi)\cos\psi, \qquad (7.6.15)$$

$$\psi(t) = t + \varphi(t). \qquad (7.6.16)$$

Note that at this stage, these coupled first-order ODEs are *exact* and, in general, cannot be solved to obtain $a(t,\epsilon)$ and $\varphi(t,\epsilon)$. However, a first approximation can be found by making the following heuristic physical argument:

(i) The right-hand sides of Equations (7.6.14) and (7.6.15) are periodic functions of ψ, with period 2π. Thus, over any interval of 2π in ψ, function, f, is bounded, then

$$\frac{da(t)}{dt} = O(\epsilon), \quad \frac{d\varphi(t)}{dt} = O(\epsilon). \qquad (7.6.17)$$

If $0 < \epsilon \ll 1$, then these derivatives will change very little on any ψ interval of 2π. Therefore, if we average the right-hand side of Equations (7.6.14) and (7.6.15) over the interval 2π in ψ, then the resulting ODEs may be a useful first approximation for calculating $a(t,\epsilon)$ and $\varphi(t,\epsilon)$. Doing this gives

$$\frac{da(t)}{dt} = -\left(\frac{\epsilon}{2\pi}\right)\int_0^{2\pi} f(a\cos\psi, -a\sin\psi)\sin\psi\,d\psi, \qquad (7.6.18)$$

$$\frac{d\varphi(t)}{dt} = -\left[\frac{\epsilon}{2\pi a(t)}\right]\int_0^{2\pi} f(a\cos\psi, -a\sin\psi)\cos\psi\,d\psi, \qquad (7.6.19)$$

Note that the two integrals will depend only on $a(t)$, since φ is included in ψ. Therefore, these approximate equations take the form

$$\frac{da(t)}{dt} = \epsilon A(a), \quad \frac{d\varphi(t)}{dt} = \epsilon B(a), 0 < \epsilon \ll 1. \qquad (7.6.20)$$

The procedure for solving these coupled ODEs is to first solve the first ODE for $a(t,\epsilon)$, then substitute it into the right-hand side of the second equation and integrate it to obtain $\varphi(t,\epsilon)$. This is the method of first-order averaging.

7.6.2 Higher-Order Corrections

The method of averaging can be extended to higher orders in ϵ. This was done by Krylov, Bogoliuibov and Mitropolsky and is represented by the following expansions

$$x(t,\epsilon) = a \cos \psi + \epsilon u_1(a,\psi) + \epsilon^2 u_2(a,\psi) + \cdots \qquad (7.6.21)$$

$$\frac{da}{dt} = \epsilon A_1(a) + \epsilon^2 A_2(a) + \cdots, \qquad (7.6.22)$$

$$\frac{d\psi}{dt} = 1 + \epsilon B_1(a) + \epsilon^2 B_2(a). \qquad (7.6.23)$$

They construct a scheme for step-by-step calculation of the functions on the right-hand sides of these equations from a knowledge of Equation (7,6.1). This procedure is long and complicated, and also has certain difficulties that make it not useful for determining corrections to the basic first-order averaging procedure. The book by Mickens (1996) provides an example where 'spurious limit-cycles' can occur in the application of these higher-order techniques.

The author's experience is consistent with the fact that many 'standard expansion (in ϵ) methods' only provide valid approximation in the lowest order ϵ. All of the calculations that we carry out in the next subsection are done only to $O(\epsilon)$. In a fundamental sense, unless one is very careful, most of the expansion techniques will only provide qualitative information on the solutions we seek.

7.6.3 Examples

7.6.3.1 The van der Pol Oscillator

The van der Pol oscillator equation is

$$\frac{d^2 x}{dt^2} + x = \epsilon(1 - x^2)\frac{dx}{dt}. \qquad (7.6.24)$$

Two other similar nonlinear oscillators are

$$\text{Lewis equation} : \quad \frac{d^2 x}{dt^2} + x = \epsilon(1 - |x|)\frac{dx}{dt}, \qquad (7.6.25)$$

$$\text{Rayleigh equation} : \quad \frac{d^2 x}{dt^2} + x = \epsilon\left[1 - \left(\frac{i}{3}\right)\left(\frac{dx}{dt}\right)^2\right]\frac{dx}{dt}. \qquad (7.6.26)$$

For the van der Pol equation

$$f\left(x, \frac{dx}{dt}\right) = (1 - x^2)\frac{dx}{dt} \tag{7.6.27}$$

and

$$f(a\cos\psi, -a\sin\psi) = (1 - a^2\cos^2\psi)(-a\sin\psi). \tag{7.6.28}$$

Therefore,

$$\frac{da}{dt} = \left(\frac{\epsilon}{2\pi}\right)\int_0^{2\pi} a(1 - a^2\cos^2\psi)\sin^2\psi\, d\psi, \tag{7.6.29}$$

$$\frac{d\varphi}{dt} = \left(\frac{\epsilon}{2\pi}\right)\int_0^{2\pi} (1 - a^2\cos^2\psi)\sin\psi\cos\psi\, d\psi. \tag{7.6.30}$$

The second integrand is odd in ψ and therefore equal to zero; consequently, we have

$$\frac{-d\varphi}{dt} = 0 \implies \varphi = \varphi(t,\epsilon) = \varphi_0, \varphi_0 = \text{constant}. \tag{7.6.31}$$

The first-integral can be easily evaluated to give

$$\frac{da}{dt} = \epsilon\left(\frac{a}{2}\right)\left(1 - \frac{a^2}{4}\right). \tag{7.6.32}$$

If we make a change of variable,

$$a \to z = a^2, \tag{7.6.33}$$

then z satisfies the ODE

$$\frac{dz}{dt} = \epsilon z\left(1 - \frac{z}{4}\right), \tag{7.6.34}$$

which has the solution

$$z(t,\epsilon) = \frac{z_0\, e^{(\epsilon t)}}{1 + \left(\frac{z_0}{4}\right)(e^{(\epsilon t)} - 1)}, \tag{7.6.35}$$

where

$$z_0 = z(0) = [a(0)]^2 = A_0^2. \tag{7.6.36}$$

Therefore,
$$a(t,\epsilon) = \frac{A_0 e^{\frac{\epsilon t}{2}}}{\left[1 + \left(\frac{A_0^2}{4}\right)(e^{\epsilon t} - 1)\right]^{\frac{1}{2}}}, \qquad (7.6.37)$$

and the first-order averaging solution to the van der Pol equation is
$$x(t,\epsilon) = \frac{A_0 e^{\left(\frac{\epsilon t}{2}\right)} \cos(t + \varphi_0)}{\left[1 + \left(\frac{A_0^2}{4}\right)(e^{\epsilon t} - 1)\right]^{\frac{1}{2}}}. \qquad (7.6.38)$$

Inspection of this expression allows the following conclusions to be reached:

(i)
$$x(0,\epsilon) = 0. \qquad (7.6.39)$$

(ii) Large
$$t : x(t,\epsilon) \longrightarrow 2\cos(t + \varphi_0). \qquad (7.6.40)$$

(iii) Two fixed-points exist for the a-equation. They are located at
$$a(\epsilon, t) = \bar{a}_1 = 0, \quad a(\epsilon, t) = \bar{a}_2 = 2. \qquad (7.6.41)$$

A linear stability analysis shows that their fixed-points are, respectively, unstable and stable.

(iv) For this level of calculation, the period of the oscillation is 2π, i.e.,
$$T = 2\pi + O'(\epsilon). \qquad (7.6.42)$$

Figure 7.13 provides sketches of the limit-cycle behavior.

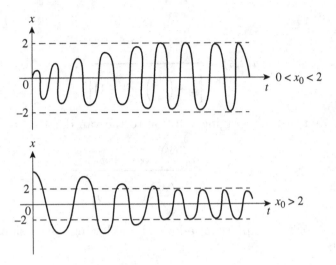

FIGURE 7.13 Sketches of $x(t, \epsilon)$ vs t for Equation (7.6.38).

7.6.3.2 Oscillator with Nonlinear Damping

The equation for a harmonic oscillator with quadratic nonlinear damping is

$$\frac{d^2x}{dt^2} + x = -\epsilon \left|\frac{dx}{dt}\right| \frac{dx}{dt}, \qquad (7.6.43)$$

for which

$$f\left(x, \frac{dx}{dt}\right) = -\left|\frac{dx}{dt}\right|\frac{dx}{dt}$$
$$\longrightarrow -|a\sin\psi|(-a\sin\psi). \qquad (7.6.44)$$

Therefore,

$$\frac{da}{dt} = -\left(\frac{\epsilon}{2\pi}\right)\int_0^{2\pi} a^2 |\sin\psi| \sin^2\psi\, d\psi, \qquad (7.6.45)$$

$$\frac{d\varphi}{dt} = -\left(\frac{\epsilon}{2\pi a}\right)\int_0^{2\pi} a^2 |\sin\psi| \sin\psi \cos\psi\, d\psi, \qquad (7.6.46)$$

where the second integral is zero because the integrand is an odd function of ψ. This means that

$$\varphi(t, \epsilon) = \varphi_0 \qquad (7.6.47)$$

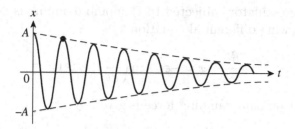

FIGURE 7.14 Sketch of x vs t, for Equation (7.6.50).

Integrating the right-hand side of Equation (7,6.45) gives

$$\frac{da}{dt} = -\left(\frac{4\epsilon}{3\pi}\right) a^2, \quad a(0) = a_0, \qquad (7.6.48)$$

whose solution is

$$a(t,\epsilon) = \frac{a_0}{\left[1 + \left(\frac{4\epsilon a_0}{3\pi}\right) t\right]} \qquad (7.6.49)$$

Therefore, the averaging method gives the following (approximate) solution for the harmonic oscillator with quadratic damping

$$x(t,\epsilon) = \frac{A \cos(t + \varphi_0)}{1 + \left(\frac{4\epsilon A}{3\pi}\right) t} \qquad (7.6.50)$$

where $A = x(0,\epsilon)$.

Note that for large times, i.e.,

$$t \gg t^* = \frac{3\pi}{4\epsilon A}, \qquad (7.6.51)$$

the solution has the behavior

$$x(t,\epsilon) \underset{\text{large } t}{\to} \left(\frac{3\pi}{4\epsilon}\right) \left[\frac{\cos(t + \varphi_0)}{t}\right]. \qquad (7.6.52)$$

See Figure 7.14 for a sketch of $x(t,\epsilon)$ vs t.

7.6.3.3 Oscillator with Coulomb Damping

The so-called 'dry friction force' is often named the Coulomb friction force or Coulomb damping force. In the context of oscillators,

214 ■ Introduction to Qualitative Methods for Differential Equations

a harmonic oscillator subjected to Coulomb damping is represented by the following differential equation

$$\frac{d^2x}{dt^2} + x = -\epsilon C \operatorname{sgn}\left(\frac{dx}{dt}\right), \quad C > 0, \tag{7.6.53}$$

where the Coulomb damping force is given by

$$F_C\left(\frac{dx}{dt}\right) = -\epsilon\, C \operatorname{sgn}\left(\frac{dx}{dt}\right), \tag{7.6.54}$$

with

$$\operatorname{sgn}(z) = \begin{cases} +1, & z > 0; \\ -1, & z < 0. \end{cases} \tag{7.6.55}$$

Note that

$$f\left(x, \frac{dx}{dt}\right) = -C \operatorname{sgn}\left(\frac{dx}{dt}\right)$$
$$\to C \operatorname{sgn}(-a \sin \psi) = -C \operatorname{sgn}(a \sin \psi)$$
$$= C \operatorname{sgn}(\sin \psi). \tag{7.6.56}$$

The last relationship follows from the fact that for oscillatory problems, the amplitude can always be selected or defined to be non-negative.

The equation for $d\varphi/dt$ is

$$\frac{d\varphi}{dt} = 0 \to \varphi(t, \epsilon) = \varphi_0 \tag{7.6.57}$$

because the corresponding integrand is odd in ψ. For da/dt, we have

$$\begin{aligned}
\frac{da}{dt} &= -\left(\frac{\epsilon}{2\pi}\right) \int_0^{2\pi} [C \operatorname{sgn}(-a \sin \psi)] \sin \psi \, d\psi \\
&= -\left(\frac{\epsilon C}{2\pi}\right) \int_0^{2\pi} \operatorname{sgn}(\sin \psi) \cdot \sin \psi \, d\psi \\
&= -\left(\frac{\epsilon C}{2\pi}\right) \left[\int_0^{\pi} \sin \psi \, d\psi - \int_{\pi}^{2\pi} \sin \psi \, d\psi\right] \\
&= -\left(\frac{2\epsilon C}{\pi}\right).
\end{aligned} \tag{7.6.58}$$

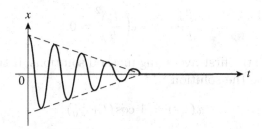

FIGURE 7.15 $x(t)$ vs t for the Coulomb damped harmonic oscillator $A = a_0$.

Or, in more detail

$$\frac{da}{dt} = \begin{cases} -\left(\frac{2\epsilon C}{\pi}\right), & a > 0, \\ 0, & a = 0. \end{cases} \qquad (7.6.59)$$

Integrating this last equation gives for $a(t, \epsilon)$, the expression

$$a(t, \epsilon) = \begin{cases} a_0 - \left(\frac{2\epsilon c}{\pi}\right)t, & 0 < t \leq t^*; \\ 0, & t > t^*, \end{cases} \qquad (7.6.60)$$

for

$$a_0 = x(0, \epsilon) > 0, \qquad t^* = \frac{\pi a_0}{2\epsilon C}. \qquad (7.6.61)$$

Thus, the damped oscillating will only exist for a finite time, $t = t^*$. Further, the total number of oscillations will be finite. If we denote this number by N, then

$$N \simeq \frac{t^*}{2\pi} = \frac{a_0}{4\epsilon C}. \qquad (7.6.62)$$

Figure 7.15 is a sketch of the function

$$x(t, \epsilon) = a(t, \epsilon) \cos(t + \varphi_0), \qquad (7.6.63)$$

where $a(t, \epsilon)$ is the results expressed in Equation (7.6.60).

7.6.3.4 Oscillators Having Quadratic Terms

Two interesting examples to examine are harmonic oscillators with the addition of the quadratic terms x^2 and $(dx/dt)^2$, i.e.,

$$\frac{d^2 x}{dt^2} + x + \epsilon x^2 = 0, \qquad (7.6.64)$$

$$\frac{d^2x}{dt^2} + x + \epsilon\left(\frac{dx}{dt}\right)^2 = 0. \qquad (7.6.65)$$

If we calculate the first-averaging method solutions, it turns out that they both have the solution

$$x(t,\epsilon) = A\cos(t+\varphi_0), \qquad (7.6.66)$$

where A and φ_0 are constant. This means that to the first approximation in the (small) parameter, ϵ, no change occurs in the behavior of the basic solution to the harmonic oscillator if quadratic nonlinear terms are included.

Figures 7.16, 7.17 and 7.18 give the details of the $x-y = dx/dt$ phase-space, with sketches of typical trajectories, for $\epsilon > 0$. We have also included the third quadratic nonlinear term equation, i.e.,

$$\frac{d^2x}{dt^2} + x + \epsilon x \frac{dx}{dt} = 0. \qquad (7.6.67)$$

Observe that all three cases produce entirely different phase-space structures.

We now provide more details on these three quadratically modified harmonic oscillator ODEs.

$$\frac{d^2x}{dt^2} + x + \epsilon x^2 = 0$$

The first-order, coupled system ODEs are

$$\frac{dx}{dt} = y, \quad \frac{dy}{dt} = -x(1+\epsilon x). \qquad (7.6.68)$$

There are two fixed-points located at

$$\left(\bar{x}^{(1)}, \bar{y}^{(1)}\right) = (0,0), \quad \left(\bar{x}^{(2)}, \bar{y}^{(2)}\right) = \left(-\frac{1}{\epsilon}, 0\right). \qquad (7.6.69)$$

The ODE for the trajectories is

$$\frac{dy}{dx} = -\frac{x(1+\epsilon x)}{y}. \qquad (7.6.70)$$

The nullclines are

$$\frac{dy}{dx} = 0 : \begin{cases} \text{along the } y\text{-axis,} \\ \text{along } y = (-1/\epsilon). \end{cases} \qquad (7.6.71)$$

Applications ■ 217

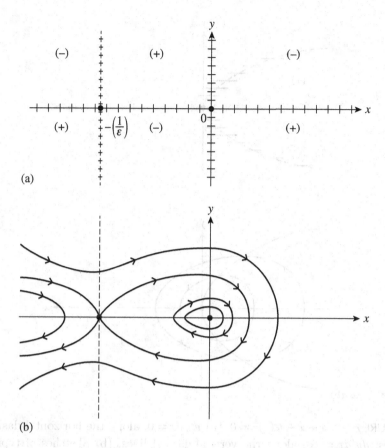

FIGURE 7.16 $\ddot{X} + x + \epsilon x^2 = 0$. (a) $dy/dx = 0$, along the horizontal-dashed lines; $dy/dx = \infty$, along the vertical-dashed lines. (b) Sketches of typical trajectories in the x–y phase-space.

$$\frac{dy}{dx} = \infty \ : \ \text{along the } x\text{-axis.} \tag{7.6.72}$$

Since Equation (7.6.70) is invariant under

$$y \to -y, \tag{7.6.73}$$

the solutions are invariant in reflection on the x-axis.

The following conclusions can be reached from an inspection of Figure 7.16:

(a) Our nonlinear oscillator has both periodic solutions (closed curves) and nonperiodic, unbounded solutions.

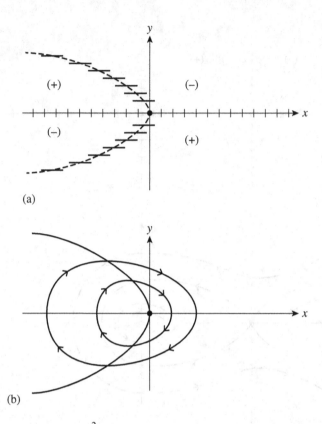

FIGURE 7.17 $\ddot{x} + x + t\epsilon(\dot{x})^2 = 0$. (a) $dy/dx = 0$, along the horizontal-dashed lines; $dy/dx = \infty$, along the vertical-dashed lines. (b) Sketches of typical trajectories in the $x - y$ phase-space.

(b) The periodic solutions take place around the fixed-point, $\left(\bar{x}^{(1)}, \bar{x}^{(2)}\right) = (0, 0)$.

(c) The fixed-points are of the following types:

$$\left(\bar{x}^{(1)}, \bar{x}^{(1)}\right) = (0, 0) \; : \; \text{center}, \tag{7.6.74}$$

$$\left(\bar{x}^{(2)}, \bar{x}^{(2)}\right) = \left(-\frac{1}{\epsilon}, 0\right) : \text{hyperbolic (saddle point).} \tag{7.6.75}$$

(d) There is a homoclinic orbit that begins and ends at the same fixed-point, $\left(\bar{x}^{(2)}, \bar{y}^{(2)}\right)$. Trajectories inside it are periodic; those lying outside it have unbounded trajectories.

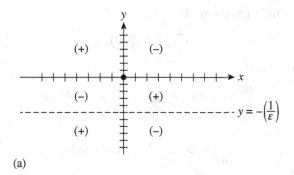

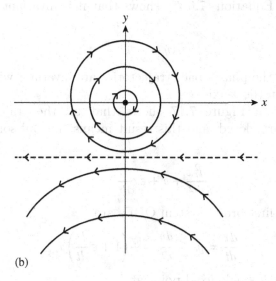

FIGURE 7.18 $\ddot{x} + x + \epsilon x \dot{x} = 0$. (a) $dy/dx = 0$, along the horizontal-dashed lines; $dy/dx = \infty$, along the vertical-dashed lines. (b) Sketches of typical trajectories in the $x - y$ phase-space.

$$\frac{d^2x}{dt^2} + x + \epsilon\left(\frac{dx}{dt}\right)^2 = 0.$$

The system form of this ODE is

$$\frac{dx}{dt} = y, \quad \frac{dy}{dt} = -x - \epsilon y^2, \qquad (7.6.76)$$

and the corresponding ODE for the trajectory curves is

$$\frac{dy}{dx} = -\left(\frac{x + \epsilon y^2}{y}\right). \qquad (7.6.77)$$

There is a single fixed-point
$$(\bar{x}, \bar{y}) = (0, 0). \tag{7.6.78}$$

The nullclines are
$$\frac{dy}{dx} = 0: \text{ along the curve } x + \epsilon y^2 = 0, \tag{7.6.79}$$

$$\frac{dy}{dx} = \infty : \text{ along the } x\text{-axis}. \tag{7.6.80}$$

Inspection of Equation (7.6.77) shows that it is invariant under the transformation
$$y \to -y, \tag{7.6.81}$$

consequently, the phase-space trajectories are invariant with respect to reflection in the x-axis.

Inspection of Figure 7.17 shows that all the trajectories in phase-space are closed, and this result means that all solutions are periodic

$$\frac{d^2x}{dt^2} + x + \epsilon x \frac{dx}{dt} = 0.$$

The coupled, first-order, system ODEs are
$$\frac{dx}{dt} = y, \quad \frac{dy}{dt} = -\left(1 + \epsilon \frac{dx}{dt}\right) x, \tag{7.6.82}$$

and they have a single fixed-point at
$$(\bar{x}, \bar{y}) = (0, 0). \tag{7.6.83}$$

The ODE determining the trajectories in phase-space is
$$\frac{dy}{dx} = -\left(\frac{1 + \epsilon y}{y}\right) x. \tag{7.6.84}$$

The nullclines are
$$\frac{dy}{dx} = 0 : \begin{cases} \text{along the } y\text{-axis;} \\ \text{along } y = -\left(\frac{1}{\epsilon}\right). \end{cases} \tag{7.6.85}$$

$$\frac{dy}{dx} = \infty : \text{ along the } x\text{-axis}. \tag{7.6.86}$$

From Figure 7.19, the following conclusion can be reached:

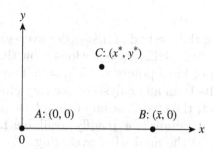

FIGURE 7.19 Minimum number of fixed-points for a realistic predator–prey system.

(a) There are three classes of solutions.

Periodic solutions for trajectories having

$$x_0 = x(0), \text{arbitrary},$$

$$y_0 = y(0) > -\left(\frac{1}{\epsilon}\right).$$

If $y_0 = y(0) = -\left(\frac{1}{\epsilon}\right)$, then the solution is

$$x(t) = x_0 - t.$$

Unbounded solutions for any point for which the starting conditions are

$$x_0 = x(0), \text{arbitrary},$$

$$y_0 = y(0) < -\left(\frac{1}{\epsilon}\right).$$

(b) The flow of the phase-space trajectories has the appearance of a flowing fluid.

7.6.4 Summary

Our presentation of the method of first-order averaging and its application to several 1-dim oscillatory equations show that it is valid only in a neighborhood of fixed-points at $(\bar{x}, \bar{y}) = 0$. This result is consistent with the results from an analysis of the trajectories in the 2-dim phase-plane. In fact, the examination of the phase-plane trajectories may give important details not readily available from the explicit formulas provided by the method of averaging.

7.7 CULLING IN PREDATOR–PREY SYSTEMS

Culling of animals is a process of selectively removing animals from a population to control their numbers such that a particular set of goals are achieved. The removal methods may include killings, isolation, etc. The reasons for culling are varied and include some or all of the following outcomes:

- Economic benefits
- Disease reduction and control
- Achieving sustainable animal populations
- Elimination of animals that pose a serious threat to human health and activities

However, culling can have negative consequences such as

- Harming unrelated animal species
- Destroy the biodiversity of a given ecosystem
- Drive the original animal species to extinction

Our interest in the issue of predator culling in predator–prey systems had its genesis in a news story by Goldfarb (2016). He discussed evidence to suggest that the outcomes of predator culling may not be consistent with prior notions of what should occur, i.e., culling should end or at least cause a reduction in predation. Goldfarb's article was a commentary and analysis of a research article by Prentice et al. (2019). In the Abstract, they state

> Culling wildlife to control disease can lead to decreases and increases in disease levels, with apparently conflicting respones observed, even for the same wildlife-disease

system .... We show that if population reduction is too low, or too few groups are targeted, a perturbation effect' is observed, whereby culling leads to increased movement and disease spread. We also demonstrate the importance of culling across appropriate time scales, with otherwise successful control strategies leading to increased disease if they are not implemented for long enough.

The main purpose of this section is to present the work of Mickens and his collaborators on culling of predators in simple predator–prey mathematical models based on two coupled ODEs. We give arguments to show that culling does not change the long-term behavior of the population dynamics of either the prey or predator populations.

The following assumptions are assumed to hold:

(i) Prey can exist without the predator and generally have a non-negative, net birth rate;
(ii) Predators consume the prey as their only food source;
(iii) All environmental factors are assumed to be constant;
(iv) Neither population is eradicated by the culling.

7.7.1 Predator–Prey Models

A broad variety of predator–prey mathematical models have been constructed to analyze these types of two interacting population systems. The references, to be listed, provide information on the history of these models, why they were created, and details on how they were constructed:

(a) R. M. May, *Science*, Vol. 177 (1972), 900–902.
(b) R. M. May, *Stability and Complexity in Model Ecosystems*, 2nd Edition (Princeton University Press, Princeton, 1974), see Chapter 4.
(c) J. D. Murray, *Mathematical Biology* (SpringerVerlag, Berlin, 1989).
(d) M. Rosensweig, *Science*, Vol. 171 (1971), 385–387.

Representative samplings of these models are:
Lotka–Volterra

$$\frac{dx}{dt} = x(a - by), \tag{7.7.1}$$

$$\frac{dy}{dt} = y(-c + dx). \tag{7.7.2}$$

Verhulst–Pearl

$$\frac{dx}{dt} = rx(K - x) - bxy, \tag{7.7.3}$$

$$\frac{dy}{dt} = y(-c + dxy). \tag{7.7.4}$$

Gompertz

$$\frac{dx}{dt} = (rx)\,\mathrm{Ln}\left(\frac{K}{x}\right) = -bxy, \tag{7.7.5}$$

$$\frac{dy}{dt} = y(-c + dx). \tag{7.7.6}$$

Logistic–Ivlev

$$\frac{dx}{dt} = (rx)(K - x) - y[1 - \exp(-bx)], \tag{7.7.7}$$

$$\frac{dy}{dt} = y\{-c + d[1 - \exp(-bx)]\}. \tag{7.7.8}$$

Logistic–Holling–Leslie

$$\frac{dx}{dt} = (rx)(K - x) - \frac{axy}{b + x}, \tag{7.7.9}$$

$$\frac{dy}{dt} = cy\left(1 - \frac{y}{gx}\right). \tag{7.7.10}$$

In these equations, the parameters, (a, b, c, d, r, g, k) are non-negative and, $x = x(t)$ and $y = y(t)$, denote, respectively, the prey and predator populations at time t.

7.7.2 General Properties of Predator–Prey Models

There are a few rules that can be used to guide the construction of mathematical models for predator–prey systems (PPS); the three most important are:

In the absence of the predator, the prey population should be bounded. Note that this requirement eliminates the standard, well-liked, Lotka–Volterra model.

The mathematical structure of the two, coupled, nonlinear ODEs should be

$$\frac{dx}{dt} = xF(x,y), \quad \frac{dy}{dt} = yG(x,y). \qquad (7.7.11)$$

In the $x - y$ phase-plane, the ODE system, given in Equation (7.7.11) must have at least three fixed-points or equilibrium solutions. See Figure 7.19 for more details.

The three fixed-points must have the following properties:

(i) Fixed-point A : $(0,0)$ corresponds to a state where there are no predators or prey. It must be *unstable*. The reason for this to be true is that the introduction of a small amount of prey will cause, in the absence of predators, an increase in the prey population.

(ii) The fixed-point B : $(\bar{x}, 0)$ is the equilibrium state of the prey population in the absence of a predator population. It must also be *unstable* since the introduction of a small number of predators will decrease the prey population from its equilibrium value.

(iii) The fixed-point C : (x^*, y^*) denotes a fixed-point that indicates an equilibrium state in which the prey and predator populations can co-exist. It must obviously be stable.

In general, the fixed-points at A : (00) and B : $(\bar{x}, 0)$ are staddle-points. However, the fixed-point at C : (x^*, y^*) can *a priori* be of three types:

- A (linear or nonlinear) center, which has neutral stability;
- A stable node;
- An unstable node.

Figure 7.20 illustrates these three possibilities. Note that the case for a center is not ecological realistic since the initial conditions can be selected such that the prey and predator oscillation amplitudes can be made arbitrarily large. This implies that the Lotka–Volterra model, while widely used to illustrate predator–prey dynamics, is not realistic.

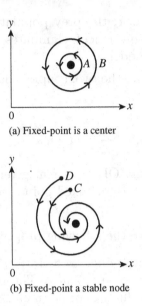

(a) Fixed-point is a center

(b) Fixed-point a stable node

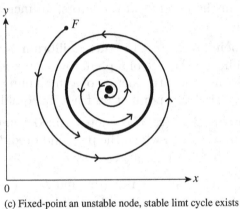

(c) Fixed-point an unstable node, stable limt cycle exists

FIGURE 7.20 The dot represents the fixed-point at (x^*, y^*). There are three possibilities for the stability of this fixed-point as indicated in the diagrams. In (c), the continuous closed curve is a limit-cycle.

If (x^*, y^*) corresponds to a stable node, then the near-by trajectories all spiral into it, as shown in Figure 7.20(b). However, the third case is the most interesting situation; see Figure 7.20(c). Here, we have an unstable node, enclosed by a stable limit-cycle, which means that essentially all initial conditions give rise to trajectories approaching the limit-cycle. For this case, the minimum and maximum values of the prey and predator populations, along with the period of the osculation, are determined by the parameters appearing in the mathematical model and not by the initial conditions.

7.7.3 Culling

Within the context of a mathematical model for the interacting dynamics of a particular sat of predator and prey populations, we define culling of the predator population as follows:

Definition
Culling is the (external) reduction of the predator population at time, $t = t_0$, by an amount y_c, i.e.,

$$y(t_0) \to y(t_0) - y_c, \qquad (7.7.12)$$

where

$$0 < y_c < y(t_0). \qquad (7.7.13)$$

Generally, culling is done with the expectation that there will be an eventual increase in the prey population. Further, the culling action may be repeated at later times at either regular or irregular time intervals.

Here, our main purpose is to provide arguments to show, within the context of standard mathematical models of predator–prey population dynamics, that culling does not change the long-term population numbers of either the prey or predator. We do this by examining what happens with just one culling event.

7.7.4 Culling the Predator

Culling the predator at time, t_0, is equivalent to moving from the phase-plane point, $(x(t_0); y(t_0))$, to $(x(t_0), y(t_0) - y_c)$, where $y_c > 0$ is the magnitude of the cull.

In Figures 7.20 and 7.21 , in the diagrams, '1' denotes the location of the trajectory at time t_0, i.e., $(x(t_0), y(t_0))$, while '2' is the position in the phase-plane after the culling has taken place, $(x(t_0), y(t_0) - y_0)$. The results from these diagrams are a visual demostration that with a single cull the long-term dynamics does

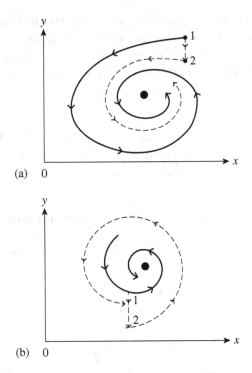

FIGURE 7.21 The fixed-point, (x^*, y^*) is a stable node. The full line is the original trajectory; the dashed line is the perturbed (after culling) path.

not change, i.e., the predator and prey populations return to either the fixed-point, (x^*, y^*), or to the surrounding limit-cycle (Figure 7.22).

7.7.5 Summary

The results of this section imply that within the framework of standard, deterministic, ODE-based mathematical models for one prey and one predator, culling of the predator does not have long-term affects on the prey or predator populations. Consequently, there are good reasons to investigate other types of mathematical models, such as ones that include stochastic effects. Careful inspection of Figures 7.20 and 7.21 shows that culling of the predator population may in some instances even cause, later on, an increase in the predator population.

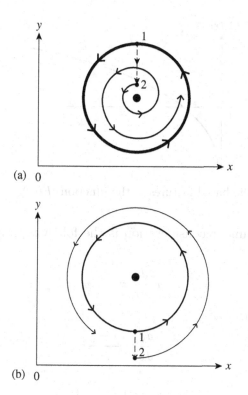

FIGURE 7.22 The fixed-point (x^*, y^*) is an unstable node and is inside a stable limit-cycle. After culling, light lines, the trajectories still approach the limit-cycle.

Finally, we should be aware that all the considerations of the section were arrived at without the need to solve any equations. This is the essence of the qualitative analysis methodology.

7.8 A LINEAR ODE: $Y' = (X - Y)/X^2$

One of the most frustrating aspects involved with the mathematical modeling of a physical system is finding out that the differential equation has an exact solution, but this solution contains functions that you have never encountered before. In this section, we will illustrate this phenomena and show one possibility as to how to deal with it.

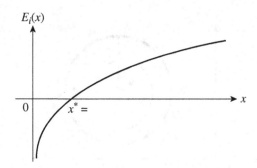

FIGURE 7.23 The based features of the function, $Ei(x)$.

The ODE under consideration is the following first-order, linear equation

$$x^2 \frac{dy}{dx} = x - y, \qquad (7.8.1)$$

which we rewrite as

$$\frac{dy}{dx} = \frac{x-y}{x^2} \qquad (7.8.2)$$

or

$$\frac{dy}{dx} + \left(\frac{1}{x^2}\right) y = \frac{1}{x}. \qquad (7.8.3)$$

There are several methods of solving this equation for its 'exact' solution. For example, using any one of many differential equation solvers, the following answer is given

$$y(x) = -Ei\left(-\frac{1}{x}\right) \cdot e^{\frac{1}{x}} + c_i e^{\frac{1}{x}}, \qquad (7.8.4)$$

and it is stated that $Ei(x)$ is the so-called exponential integral

$$Ei(x) = \int_{-\infty}^{x} \frac{e^t}{t} dt. \qquad (7.8.5)$$

The general features of $Ei(x)$ are sketched is Figure 7.23. Note that

$$Ei(0) = -\infty; \quad Ei(x^*) = 0; \quad x^* = 0.3725.... \qquad (7.8.6)$$

Few physical scientists will recognize this function and so having it definition, Equation (7.8.16), will be of little value. One has to consult either books on 'special functions' or use numerical methods to obtain useful information.

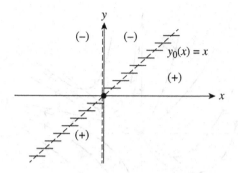

FIGURE 7.24 Nullclines for $y' = (x-y)/x^2$. Note that the nullclines divide the $x-y$ plane into four regions, each having a definite sign for $y'(x)$.

In the remainder of this section, we will show how a useful approximate solution can be calculated. However, before doing this, an analysis of the general features of 'E', Equation (7.8.13) will be done.

7.8.1 Qualitative Analysis

The first thing we will do is to reinterpret Equation (7.8.13) as the ODE that gives the trajectories in an $x-y$ planar phase-space. Therefore, this 2-dim dynamic system can be represented by the following coupled ODEs

$$\frac{dx}{dz} = x^2, \quad \frac{dy}{dz} = x - y, \tag{7.8.7}$$

where a dummy independent variable, z, has been introduced. Since

$$\frac{dy}{dx} = \frac{dy/dt}{dx/dt}$$
$$= \frac{x-y}{x^2}, \tag{7.8.8}$$

it follows that there is only one fixed-point and it is located at the origin, i.e.,

$$(\bar{x}, \bar{y}) = (0, 0). \tag{7.8.9}$$

In this $x-y$ plane, the nullclines are

$$\frac{dy}{dx} = 0: \text{ points on the curve} y_0(x) = x. \tag{7.8.10}$$

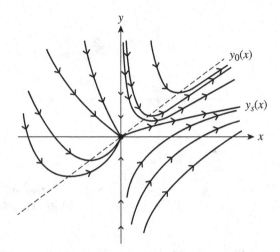

FIGURE 7.25 Typical trajectories in the $x-y$ plane. These trajectories are the solutions we are seeking. The dashed line is the zero nullcline.

$$\frac{dy}{dx} = \infty: \text{ points on the } y\text{-axis}. \qquad (7.8.11)$$

With this information, Figures 7.23 gives the nullclines and Figure 7.25 sketches representations of the general flow of the trajectories.

We now summarize some of the important features of the trajectory flows as presented in Figure 7.25:

(i) The neighborhood of the only fixed-point, at $(\bar{x}, \bar{y}) = (0, 0)$, is complex.

(ii) If we restrict our analysis to $x > 0$, then eventually all trajectories become positive and monotonically increase.

(iii) In the first quadrant, trajectories that initially start above the line, $y_0(x) = x$, i.e., the zero nullcline, decrease until they cross $y_0(x)$ with zero slope and then increase smoothly.

(iv) There is a curve, denoted by $y_s(x)$, for which all the trajectories in (iii) lie above it. Further, all trajectories that originate in the fourth quadrant lie below it.

(v) For $x \geq 0$, we have

$$\mathop{\text{Lim}}_{x \to 0^+} y(x) = +\infty, \quad if \ y > 0; \text{first quadrant}.$$

$$\text{Lim}_{x\to 0^+} y(x) = -\infty, \quad \text{if } y < 0; \text{fourth quadrant}.$$

Finally, observe that the real value of considering our ODE as the trajectory determining equation for a 2-dim system is that this procedure allows us to construct a phase-plane representation that gives us information on the solutions, $y(x)$, of the original ODE. Note that the dummy varable, z, never appears.

7.8.2 Construction of an Approximate Solution

We will now use the method of dominant balance to help construct an approximation to the exact solution of the ODE

$$\frac{dy}{dx} = \frac{1}{x} - \left(\frac{1}{x^2}\right) y. \qquad (7.8.12)$$

For x large, 'assume' that the second term on the right-hand side is negligible to react to the first term, i.e.,

$$\text{Large} x \ : \ \frac{1}{x} \gg \frac{y}{x^2}. \qquad (7.8.13)$$

Therefore, our ODE can be approximated by

$$\frac{dy}{dx} \simeq \frac{1}{x}, \qquad (7.8.14)$$

which has the solution

$$y_L(x) = \text{Ln}(|x|) + C_1, \qquad (7.8.15)$$

where C_1 is a constant. Note that under this condition, the inequality in Equation (7.8.13) holds.

Now, assume that for small x, i.e., in the neighborhood of 0^+, the second term on the right-hand side of Equation (7.8.12) dominates. Under this condition our 'approximation' ODE is

$$\frac{dy}{dx} \simeq -\left(\frac{1}{x^2}\right) y \qquad (7.8.16)$$

with solution

$$\text{Ln}(y) = \frac{1}{x} + C_2, \qquad (7.8.17)$$

and this can be rewritten to the form

$$y_S(x) = C_3 e^{\frac{1}{x}}. \tag{7.8.18}$$

We will take as our approximation to the exact solution the following expression

$$y_{\text{app}}(x) = y_L(x) + y_S(x) \tag{7.8.19}$$

$$= \text{Ln}(|x|) + C_3 e^{\frac{1}{x}} + C_1.$$

Since our original ODE is of first order, only one arbitrary constant should appear. Therefore, we take

$$C_1 = 0. \tag{7.8.20}$$

This can also be justified by observing that in Equation (7.8.15),

$$x\text{large}: \text{Ln}(|x|) \gg +C_1. \tag{7.8.21}$$

Finally, we obtain the following result for an approximation to the exact solution of Equation (7.8.12),

$$y_{\text{app}}(x) = \text{Ln}(|x|) + Ce^{\frac{1}{x}}, x > 0, \tag{7.8.22}$$

where C is an arbitrary constant.

If Equation (7.8.12) satisfies the initial conditions

$$y_0 = y(x_0), \quad x_0 > 0, \tag{7.8.23}$$

and if we impose this requirement on $y_{\text{app}}(x)$, then

$$y_0 = \text{Ln}(|x_0|) + Ce^{\frac{1}{x_0}}, \tag{7.8.24}$$

then solving for C gives

$$C = [y_0 - \text{Ln}(|x_0|)] e^{-\frac{1}{x_0}}. \tag{7.8.25}$$

Finally, substituting this into Equation (7.8.22) gives

$$y_{\text{app}}(x) = \text{Ln}(|x|) + [y_0 - \text{Ln}(|x_0|)] \cdot \exp\left(\frac{x_0 - x}{x_0 x}\right). \tag{7.8.26}$$

Now, whether this is a good approximation to the actual exact solution or not depends on what the user of this relation needs.

Taking the first derivative of

$$y_{\text{app}}(x) = \text{Ln}(|x|) + c \exp\left(\frac{1}{x}\right) \qquad (7.8.27)$$

gives

$$\frac{dy_{\text{app}}(x)}{dx} = \frac{1}{x} - \left(\frac{c}{x^2}\right) e^{\frac{1}{x}}. (7,8.17)$$

Also,

$$ce^{\left(\frac{1}{x}\right)} = y_{\text{app}}(x) - \text{Ln}(|x|); (7, 8.18)$$

therefore,

$$\frac{dy_{\text{app}}(x)}{dx} = \frac{1}{x} - \frac{y_{\text{app}}(x)}{x^2} + \frac{\text{Ln}(|x|)}{x^2}. \qquad (7.8.28)$$

Comparing the ODEs in Equations (7.8.12) and (7.8.28), we see that the ODE for $y_{\text{app}}(x)$ has an extra term on its right-hand side, and this term goes to zero as $x \to \infty$. Consequently, it should be expected that $y_{\text{app}}(x)$ is a good approximation only for large x. Our 'numerical experimental' confirms this expectation.

7.8.3 Summary

The work of this section provides an example of an ODE whose solution is given in terms of one of the non-elementary functions, namely, the exponential integral function, $E_i(x)$. This function is generally not known to most scientists and engineers. To aid in understanding the solution to the ODE, Equation (7.8.12), we first studied the major features of the solutions in the 2-dim $x - y$ phase-space. We did this under the useful construction of a dynamic system where x and y are the independent variables of the system, using a 'dummy variable', Z, Next, we constructed an approximation to the exact solution and briefly discussed it in relationship to the exact ODE.

Overall, this example illustrates the fact that having an exact analytical solution may not provide the insights needed to understand a system represented by the modeling ODE.

7.9 APPROXIMATING '1' AND '0'

7.9.1 Introduction

For most scientists, engineers and (all) mathematicians, the idea of approximating '1' or '0' is nonsense. However, we will demonstrate in this section that these concepts are valid, at least within the framework of constructing numerical schemes for differential equations and expansion-type solutions for analytical (perturbative) solutions also of differential equations.

One technique used to achieve this task is the parameter expansion method; see Senator and Bapat (1993) and Mickens (1999, 2010). A broad overview of this methodology, along with applications, is the article of He (2006):

> J. H. He, Some asymptotic methods for strongly nonlinear equations, *International Journal of Modern Physics*, Vol. 20B (2006), 1141–1199.

The following is a summary of the basic methodology of parameter expansion:

(1) First, introduce a parameter p, where

$$0 \leq p \leq 1, \tag{7.9.1}$$

and place it in the differential equation such that for $p = 1$, the original differential equation is recovered.

(2) Second, expand the dependable variable and one or more of the 'constants' appearing in the differential equation in a power series in p.

(3) The new rewritten equation is now solved under the assumption that $0 < p \ll 1$, by using some perturbation method. Denote this solution by $x(p, t)$.

(4) Finally, evaluate this p-expansion result at $p = 1$. This is taken to be a 'valid' approximation to the solution of the original equation.

Note that the parameter expansion procedure can be applied to any type of mathematical equation, although, in practice, its most important uses have been to nonlinear ordinary differential equations.

The remainder of this section is devoted to illustrating the actual uses of the ideas concerned with 'approximating one and zero'.

As a final comment, we would like to indicate that all expansion methods correspond to some form of the dominant balance technique.

7.9.2 Finite Difference Discretization of a Modified Decay ODE

Consider the following modified decay ODE,

$$\frac{dx}{dt} = -\lambda\sqrt{x}, \quad x(0) = x_0 > 0, \quad (7.9.2)$$

where λ is constant and positive. This equation only has valid physical solutions if

$$x \geq 0. \quad (7.9.3)$$

Under this condition, we have

$$\frac{dx}{dt} < 0, \quad x > 0, \quad (7.9.4)$$

and $x(t)$ decreases smoothly to zero as $t \to \infty$. Just based on the above information, the plot of $x(t)$ vs t has two possibilities, as indicated in Figure 7.26:

(A) $x(t)$ decreases smoothly to zero at $x(\infty) = 0$.

(B) $x(t)$ decreases smoothly to zero at $t = t^*$, and remains zero for $t > t^*$.

Note that inspection of Equation (7.9.2) shows that it has an equilibrium or constant solution

$$x(t) = \bar{x} = 0. \quad (7.9.5)$$

The exact solution to this ODE can be calculated using the method of separation of variables; it is

$$x(t) = \begin{cases} \left(\sqrt{x_0} - 2\lambda t\right)^2, & 0 < t < t^*; \\ 0, t > t^*; & t^* = \frac{\sqrt{x_0}}{2\lambda}. \end{cases} \quad (7.9.6)$$

Therefore, case (B) holds.

Now, suppose we wish to construct a finite-difference discretization of Equation (7.9.2). A nonstandard way of processing is to do the following:

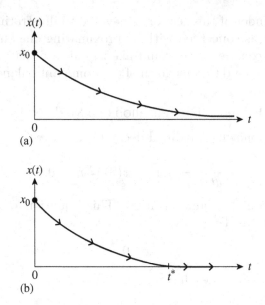

FIGURE 7.26 Possible solution behaviors for Equation (7.9.2). (a) $x(t)$ decreases smoothly to $x(\infty) = 0$. (b) $x(t)$ decreases smoothly to $x(t^*) = 0$ and remains at zero for $t > t^*$.

(i) Make the following discretization replacements

$$t \to t_k = hk, h = \Delta t, k = (0, 1, 2, ...), \quad (7.9.7)$$

$$x(t) \to x_k, \quad (7.9.8)$$

$$\frac{dx}{dt} \to \frac{x_{k+1} - x_k}{h}, \quad (7.9.9)$$

$$\sqrt{x} \to \sqrt{x_k}. \quad (7.9.10)$$

Putting these replacements in Equation (7.9.2) gives the discretization

$$\frac{x_{k+1} - x_k}{h} = -\lambda \sqrt{x_k} \quad (7.9.11)$$

or

$$x_{k+1} = x_k - (\lambda h)\sqrt{x_k}. \quad (7.9.12)$$

A closed examination of Equations (7.9.2)–(7.9.5) shows that

$$0 \leq x(t) \leq x_0. \tag{7.9.13}$$

If we now require that x_k has the properties

$$x_k \geq 0, \quad x_{k+1} \leq x_k, \tag{7.9.14}$$

for all positive values of λ and h, then the finite-difference scheme of Equations (7.9.11) or (7.9.12) may not be suitable. A way to eliminate this possibility is to return to Equation (7.9.11) and write it as

$$\frac{x_{k+1} - x_k}{h} = -\lambda \left(\frac{x_{k+1}}{\sqrt{x_k}} \right), \tag{7.9.15}$$

which can be expressed as

$$\frac{x_{k+1} - x_k}{h} = \left(-\lambda \sqrt{x_{k+1}}\right) \left(\frac{x_{k+1}}{x_k} \right)^{\frac{1}{2}}. \tag{7.9.16}$$

On comparing with Equation (7.9.2), we see that in the continuum limit

$$k \to \infty, h \to 0, \quad kh = t = \text{fixed}, \tag{7.9.17}$$

we obtain

$$\frac{dx}{dt} = -\left(\lambda \sqrt{x}\right) \cdot (1). \tag{7.9.18}$$

(Observe that $\sqrt{x_k}$ has been replaced by $\sqrt{x_{k+1}}$.) A way to interpret the second expression on the right-hand side of Equation (7.9.16) is to say that it corresponds to a discretization of '1', i.e.,

$$\text{Lim} \left(\frac{x_{(k+1)}}{x_k} \right)^{\frac{1}{2}} = 1, \tag{7.9.19}$$

where the limits are those appearing in Equation (7.9.17). A simple calculation shows that

$$\left(\frac{x_{k+1}}{x_k} \right)^{\frac{1}{2}} = 1 + O(h). \tag{7.9.20}$$

If we denote a discretization of '1' by the symbol '1_d', then for this ODE, another possibility is

$$1_d = \frac{2x_{k+1}}{x_{k+1} + x_k}. \tag{7.9.21}$$

Returning to the discretization given in Equation (7.9.15), solving for x_{k+1} provides the result

$$x_{k+1} = \left(\frac{\sqrt{x_k}}{\lambda h + \sqrt{x_k}}\right) x_k, \tag{7.9.22}$$

and we can immediately conclude that

(i) given $x_0 > 0$, then

$$x_k > 0, \quad k = (1, 2, 3, \ldots); \tag{7.9.23}$$

(ii)

$$0 \leq x_{k+1} < x_k. \tag{7.9.24}$$

(iii) for given $x_0 > 0$, then

$$\lim_{k \to \infty} x_k = 0. \tag{7.9.25}$$

Therefore, we conclude that the finite-difference discretization of Equation (7.9.2), as given by the expression in Equation (7.9.16), provides numerical solutions that are in qualitative agreement with all of the important features of the original ODE, and this is due to use of the approximation of '1' given in Equation (7.9.20).

7.9.3 $d^2x/dt^2 + x^3 = 0$

This is an example of a 'truly nonlinear osallator', see Mickens (2010). This fact implies that *NONE* of the standard perturbation can be applied to this ODE, since there is no linear term appearing in the equation and all these methods assume that the ODEs take the form

$$\frac{d^2x}{dt^2} + x = \epsilon f\left(x, \frac{dx}{dt}\right), \quad 0 < \epsilon \ll 1. \tag{7.9.26}$$

But, we know that

$$\frac{d^2x}{dt^2} + x^3 = 0 \qquad (7.9.27)$$

has all its solutions periodic. This follows easily from reformulating the ODE in terms of 2-dim phase-plane variables, i.e.,

$$\frac{dx}{dt} = y, \quad \frac{dy}{dt} = -x^3, \qquad (7.9.28)$$

from which we obtain

$$\frac{dy}{dx} = -\frac{x^3}{y}. \qquad (7.9.29)$$

Integrating gives

$$H(x, y) = \frac{y^2}{2} + \frac{x^4}{4} = \text{constant} \geq 0. \qquad (7.9.30)$$

Since

$$H(x, y) = H(x_0, y_0), \qquad (7.9.31)$$

and since the curves determined by this latter equation are simple, closed trajectories, it can be concluded that all solutions are periodic.

The p-expansion procedure for calculating approximations to the periodic solutions will now be presented.

To start, rewrite Equation (7.9.27) as

$$\ddot{x} + 0 \cdot x + px^3 = 0, \qquad (7.9.32)$$

and make the following p-expansion

$$0 = \Omega^2 + p\omega_1 + \cdots, \qquad (7.9.33)$$

$$x = x_0 + px_1 + \cdots, \qquad (7.9.34)$$

where $(\Omega^2, \omega_1, x_0, x_1)$ are to be determined and $\dot{x} = dx/dt$, etc. Substitution of these last two items into Equation (7.9.32) and collecting together terms of order p^0 and p, we find

$$p^0(\ddot{x}_0 + \Omega^2 x_0) + p(\ddot{x}_1 + \Omega^2 x_1 + \omega_1 x_0 + x_0^3) + O(p^2) = 0. \qquad (7.9.35)$$

Since this is a series in p that equal zero, its follows that each coefficient in the expansion must equal zero, i.e.,

$$\ddot{x}_0 + \Omega^2 x_0 = 0; \quad x_0(0) = A, \quad \dot{x}_0(0) = 0; \quad (7.9.36)$$

$$\ddot{x}_1 + \Omega^2 x_1 = -\omega_1 x_0 - x_0^3; \quad x_1(0) = 0, \quad \dot{x}_1(0) = 0. \quad (7.9.37)$$

The initial values for the x_0 and x_1 ODEs are consequences of the relations

$$x(0, p) = A, \quad \dot{x}(0, p) = 0. \quad (7.9.38)$$

The solution for $x_0(t)$ is

$$x_0(t) = A \cos\theta, \quad \theta = \Omega t, \quad (7.9.39)$$

where at this level of calculation, Ω is not determined. With this result, we have

$$\ddot{x}_1 + \Omega^2 x_1 = -\omega_1 A \cos\theta - (A \cos\theta)^3$$
$$= -A\left[\omega_1 + \frac{3A^2}{4}\right]\cos\theta$$
$$= -\left(\frac{A^3}{4}\right)\cos 3\theta. \quad (7.9.40)$$

The coefficient of the first term on the right-hand side of Equation (7.9.40) must equal zero if we want to have a bounded, periodic solution. The reasons for this are discussed in the book

R. E. Mickens, *Introduction to Nonlinear Oscillations* (Cambridge University Press, New York, 1981).

Therefore,

$$\omega_1 = -\left(\frac{3A^2}{4}\right), \quad (7.9.41)$$

to first order in p, we have

$$\Omega^2 = -\omega_1 p = \left(\frac{3A^2}{4}\right)p. \quad (7.9.42)$$

Therefore, $x_1(t)$ is the solution to the initial-value problem

$$\begin{cases} \ddot{x}_1 + \Omega^2 x_1 = -\left(\dfrac{A^3}{4}\right)\cos 3\theta, \\ x_1(0) = 0, \quad \dot{x}(0) = 0, \quad \theta = \Omega t = \left(\dfrac{3}{4}\right)^{\frac{1}{2}}(At). \end{cases} \qquad (7.9.43)$$

An easy, direct calculus gives

$$x_1(t) = -\left(\dfrac{A}{24}\right)(\cos\theta - \cos 3\theta) \qquad (7.9.44)$$

Consequently, to first order p, the p-parameter expansion, periodic solution for Equation (7.9.27) is, with $p = 1$, the expression

$$x(t, p = 1) = x_0(t) + x_1(t)$$
$$= A\left[\cos\theta - \left(\dfrac{1}{24}\right)(\cos\theta - \cos 3\theta)\right], \qquad (7.9.45)$$

where

$$\theta(t) = \left(\dfrac{3}{4}\right)^{\frac{1}{2}}(At). \qquad (7.9.46)$$

This result is also obtained from the use of other techniques (Mickens, 2010).

7.9.4 $\ddot{x} + x^{\frac{1}{3}} = 0$

The 'truly nonlinear oscillator'

$$\dfrac{d^2 x}{dt^2} + x^{\frac{1}{3}} = 0, \qquad (7.9.47)$$

has the system equations representation

$$\dfrac{dx}{dt} = y, \quad \dfrac{dy}{dt} = -x^{\frac{1}{3}}, \qquad (7.9.48)$$

and from this it follows that

$$\dfrac{dy}{dx} = -\dfrac{x^{\frac{1}{3}}}{y}, \qquad (7.9.49)$$

with the first-integral

$$H(x, y) = \left(\dfrac{1}{2}\right)y^2 + \left(\dfrac{3}{4}\right)x^{\frac{4}{3}} = \text{constant} > 0. \qquad (7.9.50)$$

Since the trajectories in the $x-y$ phase-plane are all closed, except for the fixed-point at $(\bar{x}, \bar{y}) = (0,0)$, all solutions to Equation (7.9.47) are periodic.

To proceed, rewrite Equation (7.9.47) as

$$x = -(\ddot{x})^3, \tag{7.9.51}$$

multiply by Ω^2, which is currently an unknown constant,

$$\Omega^2 x = -\Omega^2 (\ddot{x})^3, \tag{7.9.52}$$

and, add $\ddot{x}$ to both sides, to obtain

$$\ddot{x} + \Omega^2 x = \ddot{x} - \Omega^2 (\ddot{x})^3. \tag{7.9.53}$$

Note that this is an exact, but rearranged form of the original ODE. The p-expansion form will be taken to be

$$\ddot{x} + \Omega^2 x = p\left[\ddot{x} - \Omega^2 (\ddot{x})^3\right], \tag{7.9.54}$$

with

$$x(t) = x_0(t) + p x_1(t) + \cdots. \tag{7.9.55}$$

If Equation (7.9.55) is substituted into Equation (7.9.54), then the coefficients, respectively, of the p^0 and p terms are

$$\ddot{x}_0 + \Omega^2 x_0 = 0 \tag{7.9.56}$$

$$\ddot{x}_1 + \Omega^2 x_1 = \ddot{x}_0 - \Omega^2 (\ddot{x}_0)^3, \tag{7.9.57}$$

with initial conditions

$$\begin{cases} x_0(0) = A, & \dot{x}_0(0) = 0, \\ x_1(0) = 0, & \dot{x}_1(0) = 0. \end{cases}$$

For $x_0(t)$, we find

$$x_0(t) = A \cos\theta, \quad \theta = \Omega t, \tag{7.9.58}$$

therefore,

$$\ddot{x}_1 + \Omega^2 x_1 = \left[-\Omega^2 + \frac{3A^2\Omega^8}{4}\right] A \cos\theta \tag{7.9.59}$$

$$+\left(\frac{A^3\Omega^8}{4}\right)\cos 3\theta. \qquad (7.9.60)$$

Bounded, periodic solutions require that the coefficient of the $\cos\theta$ term be zero; this gives

$$\Omega = \left(\frac{4}{3}\right)^{\frac{1}{6}}\left(\frac{1}{A^{\frac{1}{3}}}\right) = \frac{1.0491}{A^{1/3}}. \qquad (7.9.61)$$

Calculating $X_1(t)$, with the initial conditions imposed, gives

$$X_1(t) = \left(\frac{A}{24}\right)(\cos\theta - \cos 3\theta). \qquad (7.9.62)$$

Consequently, in terms of order $-p$, the solution with $p=1$ is

$$x(t) = x_0(t) + x_1(t)$$
$$= A\left[\left(\frac{25}{24}\right)\cos\theta - \left(\frac{1}{24}\right)\cos 3\theta\right], \qquad (7.9.63)$$

with

$$\theta = \Omega t$$
$$= \left[\left(\frac{4}{3}\right)^{\frac{1}{6}}\left(\frac{1}{A^{\frac{1}{3}}}\right)\right]t. \qquad (7.9.64)$$

A measure of the accuracy of the first order in p calculation may be obtained by comparing the value of Ω in Equation (7.9.61) with that of the exact value of Ω, i.e., (Micken, 2010):

$$\Omega_{app} = \frac{1.0491}{A^{\frac{1}{3}}}, \qquad \Omega_{exact} = \frac{1.0704}{A^{\frac{1}{3}}}. \qquad (7.9.65)$$

The percentage error in our calculation is about 2\%.

7.9.5 Discussion

Like most expansion procedures, the naive use of the p-expansion method generally gives valid and/or accurate results only to first order. However, the methodology can be extended to higher orders in p, but the calculations become complex. For oscillatory equations, the correction term, $x_1(t)$, is about 4\% in magnitude of $x_0(t)$.

For purposes of discretization, the approximation of '1' method is often used to insure that some physical relevant variable is always non-negative.

COMMENTS AND REFERENCES

Section 7.1: The physical basis of Equations (7.1.1) and (7.1.2) is given in

(1) L. Dresner, *Similarity Solutions of Nonlinear Partial Differential Equations* (Pitman, Boston, 1983), see Section 4.7.

The calculations presented in this section are based on the following paper

(2) R. E. Mickens and J. E. Wilkins, Jr. Estimation of $y(0)$ for the boundary-value problem: $y'' = y^2 - zyy'$, $y'(0) = -\sqrt{3}$, $y(\infty) = 0$, *Communications in Applied Analysis*, Vol. 23 (2019), 137–146.

Also, see

(3) L. Dresner, Thermal expulsion of helium from a quenching cable-in-conduit conductor, in *Proceedings of the Ninth Symposium on the Engineering Problems of Fusion Research* (IEEE Publication No. 81CH1715-2NPS, Chicago, 1981), 26–29.

The value for $y(0) = 1.5111$ appears in the book

(4) J. D. Logan, *An Introduction to Nonlinear Differential Equations* (Wiley-Interscience, New York, 1994). See Sections 4.2 and 4.3.

Section 7.2: This section is based on the article

(5) R. E. Mickens and I. H. Herron, Approximate rational solutions to the Thomas-Fermi equation based on dynamic consistency, *Applied Mathematics Letters*, Vol. 116 (2021), 106994.

The original papers of Thomas and Fermi are

(6) L. H. Thomas, The calculation of atomic fields, *Proceedings of the Cambridge Philosophical Society*, Vol. 23 (1927), 452–548.

(7) E. Fermi, Un methodo statistico per la determinazione di elcunepriorieta dell'atomé, *Rendiconti Accademia Nazionale dei Lincei*, Vol. 6 (1927), 602–607.

Two other related, but important papers on this topic are

(8) E. Hille, On the Thomas-Fermi equation, *Proceedings of the National Academy of Sciences, USA*, Vol. 62, No. 1 (1969), 7–10.

(9) F. M. Fernández, Comments on series solutions to the Thomas-Fermi equation, *Physics Letters A*, Vol. 365 (2007), 111.

Section 7.3: This section is based on the article

(10) R. E. Mickens, Wave front behavior of traveling wave solutions for a PDE having square-root dynamics, *Mathematics and Computations in Simulations*, Vol. 82 (2012), 1271–1277.

Two interesting and (somewhat) related articles that deal with the discretization of differential equations having square-root dynamics are

(11) R. Buckmire, K. McMurtry, and R. E. Mickens, Numerical studies of a nonlinear heat equation with square root reaction term, *Numerical Methods for Partial Differential Equations*, Vol. 25 (2009), 509–518.

(12) R. E. Mickens, Exact finite difference scheme for an advection equation have square-root dynamics, *Journal of Difference Equations and Applications*, Vol. 14 (2008), 1149–1157.

Section 7.4: The methodology for the construction of the approximate TW solution is based in part on the work done in the following paper:

(13) R. Mickens and K. Oyedeji, Traveling wave solutions to modified Burgers and diffusionless Fisher PDE's, *Evolution Equations and Control Theory*, Vol. 8, No. 1 (2019), 139–147.

Section 7.4: The first part of this section is based on the following published article:

(1) R. E. Mickens and S. A. Rucker, A note on a function equation model of decay processes: Analysis and consequences, *Journal of Difference Equations and Applications* (2023), 1–6. https://doi.org/10.1080/10236198.2023.2260891.

The three publications indicated below provide an overview of the issues regarding oscillations in radioactive decay. The last article gives a listing of both experimental and theoretical papers on this topic up to 2020.

(2) T. M. Semkow et al., Oscillations in radioactive exponential decay, *Physics Letters B*, Vol. 675, No. 5 (2009), 415–419.

(3) F. Giacosa and G. Pagliara, Oscillations in the decay law: A possible quantum mechanical explanation of the anomaly in the experiment at the GSI facility, *Quantum Matter*, Vol. 2 (2013), 54–59.

(4) M. H. McDuffie et al., Anomalies in radioactive decay rates: A bibliography of measurements and theory; arXiv:2012.00153v2 [nucl-ex] (Assessed July 2, 2024).

The following should prove useful for the material given in the second part of this section.

(5) G. F. Lawler, *Random Walk and the Heat Equation* (Department of Mathematics, University of Chicago, Chicago). https://www.math.uchicago.edu/~lawler/reu.pdf (Accessed, August 1, 2024). See also: G. F. Lawler, *Random Walk and the Heat Equation* (American Mathematical Society, Providence, RI, 2010).

The results of functional equations are covered in the book

(6) A. D. Polyanin and A. V. Manzhirov, *Handbook of Integral Equations: Exact Solutions (Supplement: Some Functional Equations)* (Chapman and Hall/CRC Press, Boca Ration, FL, 2008). See also: https://eqworld.ipmnet.ru/en/solutions/fe/fe1203.pdf

Also, read the following article to understand some of the difficulties that arise when modeling with delayed PDEs.

(7) I. H. Herron and R. E. Mickens, A note on exponential-type solutions for the linear, delayed heat partial differential equation; available at: arXiv: 2006.14018v1 [math.AP] (Accessed June 24, 2020).

Section 7.5: This section is derived in large part from the article

(1) I. S. Herron and R. E. Mickens, Some exact and approximate solutions to a generalized Maxwell-Cattaneo equation, *Studies in Applied Mathematics*, Vol. 151 (7) (2023), 1–19. https://doi.org/10.1111/sapm. 12640.

Two good discussions of the heat equation and related topics are

(2) D. V. Widder, *The Heat Equation* (Academic Press, New York, 1975).

(3) D. D. Joseph, Heat waves, *Reviews of Modern Physics*, Vol. 61 (1989), 41–73; Vol. 62 (1990), 375–391.

The calculation of a 'time scale' for ODEs having the form

$$\frac{dy}{dt} = F(y)$$

is given in

(4) R. E. Mickens, *Mathematical Modelling with Differential Equations* (CRC Press, Baco Raton, FL, 2022). See Section 0.5.

Section 7.6: A general introduction to many of the techniques used to calculate approximate analytical solutions for 1-dim nonlinear ODEs having oscillatory solutions is in the following two books:

(1) A. H. Nayfeh, *Perturbation Methods* (Wiley, New York, 1973).

(2) R. E. Mickens, *Oscillations in Planar Dynamic Systems* (World Scientific, Singapore, 1996).

The early work on averaging methods for nonlinear oscillations is given in the books:

(3) N. Krylov and N. Bogoliubov, *Introduction to Nonlinear Mechanics* (Princeton University Press, Princeton, NJ, 1943).

(4) N. N. Bogoliubov and J. A. Mitropolsky, *Asymptotic Methods in the Theory of Nonlinear Oscillations* (Hindustan Publishing, New Delhi, India, 1963).

The examples given in Section 7.6.3 are from Mickens (1996), Section 3.3. A discussion of the existence of 'spurious limit-cycles' that may arise in the use of higher-order averaging techniques appears in Mickens (1996), Section 3.6.

Section 7.7: The work of this section is based on the following (unpublished) article:

(1) R. E. Mickens, M. Harlemon, and K. Oyedeji, Consequences of culling in deterministic ODE predator-prey models; arXiv: 1612.09301v2 [q-bio.PE] (Accessed January 10 2017).

This article is an expansion of the discussion presented in an e-letter to the journal *Science*, which makes comments on the following publication:

(2) B. Goldfarb, No proof that predator culls save livestock, study claims, Science, Vol. 353 (2016), 1080–1081.

which was an analysis of the paper

(3) A. Treves, M. Krofel, and J, McManus, Predator control should not be a shot in the dark, *Frontiers in Ecology and Environment*, Vol. 14, No. 7 (September 2016), 380–388.

A quick read on the ecological concept of culling is in the following two articles:

(4) Wikipedia: 'Culling', https://ent.wikipendia.org/wiki/Culling.

(5) J. C. Prentice et al., When to kill a cull: Factors affecting the success of culling wild life for disease control, *Journal of the Royal Society Interface* Vol. 16, No. 152 (March 2019).

Section 7.8: The exponential integral $Ei(x)$ appears in the exact solution to the ODE examined in this section. It is related to another exponential integral, $E_1(x)$, by means of the relation

$$Ei(x) = -E_1(-x), \quad x > 0.$$

These two exponential integrals are defined by

$$Ei(x) = \int_{-\infty}^{x} \frac{e^t}{t} dt, \quad E_1(x) = \int_{x}^{\infty} \frac{e^{-t}}{t} dt.$$

Listed below are some of the properties of these functions that are of value for our investigation:

$$\left(\frac{1}{2}\right) e^{-x} \operatorname{Ln}\left(1 + \frac{2}{x}\right) < E_1(x) < e^{-x} \operatorname{Ln}\left(1 + \frac{1}{x}\right)$$

$$E_1(x) \sim \left(\frac{e^{-x}}{x}\right)\left[1 - \frac{1}{x} + \frac{2}{x^2} + \cdots\right], \quad x \to \infty$$

$$\frac{\partial E_i(x)}{\partial x} = \frac{e^x}{x}$$

$$E_i(x) \sim -\operatorname{Ln}|x|, \quad x \to 0$$

REFERENCES TO THE EXPONENTIAL FUNCTIONS

(1) M. Abramowitz and I. Stegun, *Handbook of Mathematical Functions with Formulas Graphs, and Mathematical Tables* (Dover, New York, 1964). See Chapter 5.

(2) C. M. Bender and S. A. Orszag, *Advanced Mathematical Methods for Scientists and Engineers* (Mc Graw-Hill, New York, 1978).

(3) R. R. Sharma and B. Zohuri, A general method for accurate evaluation of exponential integrals $E_1(x)$, $x > 0$, *Journal of Computational Physics*, Vol. 25, No. 2 (1977), 199–204.

The following differential equation solver was used to obtain the 'exact' solution to the ODE:

(4) https://www.symbolab.com/solver/ordinary-differential-equation-calculator.

Some additional references to the 'method of dominant balance' are

(5) hittps://www.physicsforums.com/threads/method-of-dominant-balance.1020334.

(6) N. G. de Bruijn, *Asymptotic Methods in Analysis* (Dover, Mineola, NY, 1981).

(7) hitps://en.wikipedia.org/wiki/Method-of-dominant-balance (Accessed August 15, 2024).

Also, the book by Bender and Orszag (ref. [2]) is an excellent introduction to the method of dominant balance and includes many worked examples.

Section 7.9: The discussions and examples used to illustrate the concepts of 'approximating one and zero' are based on Chapter 4 of my book:

(1) R. E. Mickens, *Truly Nonlinear Oscillations: Harmonic Balance, Parameter Expansions, Iteration, and Averaging Methods* (World Scientific, Singapore, 2010).

The following two articles provide background to the method of parameter expansion as applied to the analysis of nonlinear oscillations:

(2) M. Senator and C. N. Bapat, A perturbations technique that works even when the non-linearity is not small, *Journal of Sound and Vibration* Vol. 164 (1993), 1–27.

(3) R. E. Mickens, Generalization of the Senator-Bapat method to systems having limit-cycles, *Journal of Sound and Vibration*, Vol. 224 (1999), 167–171.

Applications to more complex systems modeled by ODEs and PDEs are included in the references:

(4) R. Buckmire, K. McMurty, and R. E. Mickens, Numerical studies of a nonlinear heat equation with square root reaction term, *Numerical Methods for Partial Differential Equations*, Vol. 25, No. 3 (2009), 598–609.

(5) M. Chapwanya, J. M.-S. Lubuma, and R. E. Mickens, Positivity-preserving nonstandard finite difference schemes for cross-diffusion equations in biosciences, *Computers and Mathematics with Applications*, Vol. 68, No. 9 (2014), 1071–1082.

Appendix A

A.1 ALGEBRAIC RELATIONS

A.1.1 Factors and Expansions

$$(a \pm b)^2 = a^2 \pm 2ab + b^2$$
$$(a \pm b)^3 = a^3 \pm 3a^2b + 3ab^2 \pm b^3$$
$$(a + b + c)^2 = a^2 + b^2 + c^2 + 2(ab + ac + bc)$$
$$(a + b + c)^3 = a^3 + b^3 + c^3 + 3a^2(b + c)$$
$$+ 3b^2(a + c) + 3c^2(a + b) + 6abc$$
$$a^2 - b^2 = (a - b)(a + b)$$
$$a^2 + b^2 = (a + ib)(a - ib), i = \sqrt{-1}$$
$$a^3 - b^3 = (a - b)(a^2 + ab + b^2)$$
$$a^3 + b^3 = (a + b)(a^2 - ab + b^2)$$

A.1.2 Quadratic Equations

The quadratic equation

$$ax^2 + bx + c = 0$$

has the two solutions

$$x_1 = \frac{-b + \sqrt{b^2 - 4ac}}{2a},$$
$$x_2 = \frac{-b - \sqrt{b^2 - 4ac}}{2a}.$$

A.1.3 Cubic Equations

The cubic equation

$$x^3 + px^2 + qx + r = 0$$

can be reduced to

$$z^3 + az + b = 0$$

by means of the substitution

$$x = z - \frac{p}{3},$$

where

$$a = \frac{3q - p^2}{3}$$

$$b = \frac{3p^3 - 9pq + 27r}{27}.$$

With the definitions

$$A = \left[-\left(\frac{b}{2}\right) + \left(\frac{b^2}{4} + \frac{a^3}{27}\right)^{\frac{1}{2}} \right]^{\frac{1}{3}},$$

$$B = \left[-\left(\frac{b}{2}\right) - \left(\frac{b^2}{4} + \frac{a^3}{27}\right)^{\frac{1}{2}} \right]^{\frac{1}{3}},$$

the three roots of $z^3 + az + b = 0$ are

$$Z_1 = A + B,$$
$$Z_2 = -\left(\frac{A+B}{2}\right) + \sqrt{-3}\left(\frac{A-B}{2}\right),$$
$$Z_3 = -\left(\frac{A+B}{2}\right) - \sqrt{-3}\left(\frac{A-B}{2}\right).$$

Let

$$\Delta = \frac{b^2}{4} + \frac{a^3}{27},$$

then for

(i) $\Delta > 0$, there will be one real root and two complex conjugate roots.

(ii) $\Delta = 0$, there will be three real roots, for which at least two are equal.

(iii) $\Delta < 0$, there will be three real and unequal roots.

A.1.4 Expansions of Selected Functions

$$\frac{1}{1-x} = \sum_{k=0}^{\infty} x^k, \quad |x| < 1$$

$$e^x = \sum_{k=0}^{\infty} \frac{x^k}{k!}, \quad |x| < \infty$$

$$(1+x)^q = 1 + qx + \frac{q(q-1)}{2!}x^2 + \cdots + \frac{q(q-1)\cdots(q-k+1)}{k!}x^k + \cdots$$

The above series converges for $|x| < 1$.

$$\cos x = \sum_{k=0}^{\infty} (-1)^k \frac{x^{2k}}{(2k)!}$$

$$\sin x = \sum_{k=0}^{\infty} (-1)^k \frac{x^{2k+1}}{(2k+1)!}$$

$$\text{Ln}(1+x) = \sum_{k=1}^{\infty} (-1)^{k+1} \frac{x^k}{k}$$

A.2 TRIGONOMETRIC RELATIONS

A.2.1 Fundamental Properties

$$(\sin\theta)^2 + (\cos\theta)^2 = 1$$
$$-1 \leq \sin\theta \leq +1, \quad -1 \leq \cos\theta \leq 1$$
$$\sin(-\theta) = -\sin\theta, \quad \cos(-\theta) = \cos\theta$$
$$\sin(\theta + 2\pi) = \sin\theta, \quad \cos(\theta + 2\pi) = \cos\theta$$

The sine and cosine functions take the following values at the indicated angles

$$\sin(0) = 0, \quad \cos(0) = 1$$
$$\sin\left(\frac{\pi}{2}\right) = 1, \quad \cos\left(\frac{\pi}{2}\right) = 0$$
$$\sin(\pi) = 0, \quad \cos(\pi) = -1$$
$$\sin\left(\frac{3\pi}{2}\right) = -1, \quad \cos\left(\frac{3\pi}{2}\right) = 0$$

We also have

$$e^{i\theta} = \cos\theta + i\sin\theta, \quad i = \sqrt{-1}$$
$$\sin\theta = \frac{e^{i\theta} - e^{-i\theta}}{2i}$$
$$\cos\theta = \frac{e^{i\theta} + e^{-i\theta}}{2}$$

A.2.2 Sum of Angles Relations

$$\sin(\theta_1 \pm \theta_2) = \sin\theta_1 \cos\theta_2 \pm \cos\theta_1 \sin\theta_2$$
$$\cos(\theta_1 \pm \theta_2) = \cos\theta_1 \cos\theta_2 \mp \sin\theta_1 \sin\theta_2$$

A.2.3 Product and Sum Relations

$$\sin\theta_1 \cos\theta_2 = \left(\frac{1}{2}\right)[\sin(\theta_1 + \theta_2) + \sin(\theta_1 - \theta_2)]$$
$$\cos\theta_1 \sin\theta_2 = \left(\frac{1}{2}\right)[\sin(\theta_1 + \theta_2) - \sin(\theta_1 - \theta_2)]$$
$$\cos\theta_1 \cos\theta_2 = \left(\frac{1}{2}\right)[\cos(\theta_1 + \theta_2) + \cos(\theta_1 - \theta_2)]$$
$$\sin\theta_1 \sin\theta_2 = \left(\frac{1}{2}\right)[\cos(\theta_1 + \theta_2) - \cos(\theta_1 - \theta_2)]$$
$$\sin\theta_1 \pm \sin\theta_2 = 2\sin\left(\frac{\theta_1 \pm \theta_2}{2}\right)\cos\left(\frac{\theta_1 \mp \theta_2}{2}\right)$$
$$\cos\theta_1 + \cos\theta_2 = 2\cos\left(\frac{\theta_1 + \theta_2}{2}\right)\cos\left(\frac{\theta_1 - \theta_2}{2}\right)$$
$$\cos\theta_1 - \cos\theta_2 = -2\sin\left(\frac{\theta_1 + \theta_2}{2}\right)\sin\left(\frac{\theta_1 - \theta_2}{2}\right)$$

A.2.4 Derivatives and Integrals

$$\frac{d}{d\theta}\cos\theta = -\sin\theta, \quad \frac{d}{d\theta}\sin\theta = \cos\theta$$

$$\int \cos\theta\, d\theta = \sin\theta + c_1, \quad \int \sin\theta\, d\theta = -\cos\theta + c_2$$

where c_1 and c_2 are arbitrary constants.

A.2.5 Powers of Trigonometric Functions

$$(\sin\theta)^2 = \left(\frac{1}{2}\right)(1 - \cos 2\theta)$$

$$(\cos\theta)^2 = \left(\frac{1}{2}\right)(1 + \cos 2\theta)$$

$$(\sin\theta)^3 = \left(\frac{1}{4}\right)(3\sin\theta - \sin 3\theta)$$

$$(\cos\theta)^3 = \left(\frac{1}{4}\right)(3\cos\theta + \cos 3\theta)$$

A.3 HYPERBOLIC FUNCTIONS

A.3.1 Definitions and Basic Properties

$$\text{hyperbolic:} sine \sinh\theta \equiv \frac{e^\theta + e^{-\theta}}{2}$$

$$\text{hyperbolic:} cosine \cosh\theta \equiv \frac{e^\theta + e^{-\theta}}{2}$$

$$(\cosh\theta)^2 - (\sinh\theta)^2 = 1$$

$$\cosh(-\theta) = \cosh(\theta), \sinh(-\theta) = -\sinh(\theta)$$

$$\cosh(0) = 1, \quad \sinh(0) = 0$$

A.3.2 Derivatives and Integrals

$$\frac{d}{d\theta}\cosh(\theta) = \sinh(\theta), \quad \frac{d}{d\theta}\sinh(\theta) = \cosh(\theta)$$

$$\int \cosh(\theta)\, d\theta = \sinh(\theta) + c_1, \quad \int \sinh(\theta)\, d\theta = \cosh(\theta) + c_2$$

A.3.3 Other Important Relations

$$\sinh(\theta_1 \pm \theta_2) = \sinh(\theta_1)\cosh(\theta_2) \pm \cosh(\theta_1)\sinh(\theta_2)$$
$$\cosh(\theta_1 \pm \theta_2) = \cosh(\theta_1)\cosh(\theta_2) \pm \sinh(\theta_1)\sinh(\theta_2)$$
$$\sinh(\theta_1) + \sinh(\theta_2) = 2\sinh\left(\frac{\theta_1+\theta_2}{2}\right)\cosh\left(\frac{\theta_1-\theta_2}{2}\right)$$
$$\cosh(\theta_1) + \cosh(\theta_2) = 2\cosh\left(\frac{\theta_1+\theta_2}{2}\right)\cosh\left(\frac{\theta_1-\theta_2}{2}\right)$$
$$\sinh(\theta_1) - \sinh(\theta_2) = 2\cosh\left(\frac{\theta_1+\theta_2}{2}\right)\sinh\left(\frac{\theta_1-\theta_2}{2}\right)$$
$$\cosh(\theta_1) - \cosh(\theta_2) = 2\sinh\left(\frac{\theta_1+\theta_2}{2}\right)\sinh\left(\frac{\theta_1-\theta_2}{2}\right)$$

A.3.4 Relations between Trigonometric and Hyperbolic Functions

$$\sinh(\theta) = -i\sin(i\theta),\ i = \sqrt{-1}$$
$$\cosh(\theta) = \cos(i\theta)$$

A.4 IMPORTANT CALCULUS RELATIONSHIPS

A.4.1 Differentiation

Using the notation

$$f'(x) = \frac{d}{dx}f(x)$$

and (c_1, c_2) representing arbitrary constants, we have

$$\frac{d}{dx}[c_1 f(x) + c_2 g(x)] = c_1 f'(x) + c_2 g'(x)$$
$$\frac{d}{dx}[f(x)g(x)] = g(x)f'(x) + f(x)g'(x)$$
$$\frac{d}{dx}\left[\frac{f(x)}{g(x)}\right] = \frac{g(x)f'(x) - f(x)g'(x)}{[g(x)]^2}$$
$$\frac{d}{dx}[f(g(x))] = f'(g(x))g'(x)$$
$$\frac{d}{dx}e^{f(x)} = f'(x)e^{f(x)}$$

A.4.2 Integration by Parts

$$\int f(x)\,dg(x) = f(x)\,g(x) - \int g(x)\,df(x)$$

A.4.3 Differentiation of a Definite Integral

Assume $f(x,t)$ and its partial derivative $\partial f(x,t)/\partial t$ is continuous in some domain in the $x-t$ plane, which includes the rectangle

$$\Psi(t) \le x \le \varphi(t),\, t_1 \le t \le t_2,$$

where $\psi(t)$ and $\varphi(t)$ are defined and have continuous first derivatives for $t_1 \le t \le t_2$. Then for $t_1 \le t \le t_2$, we have

$$\frac{d}{dt}\int_{\psi(t)}^{\varphi(t)} f(x,t)\,dx = f[\varphi(t),t]\frac{d\varphi}{dt}$$

$$- f[\psi(t),t]\frac{d\psi}{dt} + \int_{\psi(t)}^{\varphi(t)} \frac{\partial}{\partial t}f(x,t)\,dx.$$

A.5 EVEN AND ODD FUNCTIONS

Let the real functions $f(x)$ and $g(x)$ be defined on a symmetric interval, $(-a, a)$, where a might be unbounded.

(i) A function $f(x)$ is an even function on this interval if and only if

$$f(-x) = f(x).$$

(ii) A function $f(x)$ is an odd function on this interval if and only if

$$f(-x) = -f(x).$$

(iii) Given an arbitrary function $g(x)$, defined on this interval, then it can also be written as where

$$g(x) = g^{(+)}(x) + g^{(-)}(x),$$

$$g^{(+)}(x) = \frac{g(x) + g(-x)}{2},$$

$$g^{(-)}(x) = \frac{g(x) - g(-x)}{2}.$$

The functions $g^{(+)}(x)$ and $g^{(-)}(x)$ are, respectively, the even and odd parts of $g(x)$.

(iv) Let $f(x)$ and $g(x)$ be both even functions or both odd functions, then $h(x) = f(x)g(x)$ is an even function.

(v) If $f(x)$ is an even function and $g(x)$ is an odd function, the $h(x) = f(x)g(x)$ is an odd function.

(vi) Let $f(x)$ be an even function, then

$$\int_{-a}^{a} f(x)\,dx = 2\int_{0}^{a} f(x)\,dx.$$

(vii) Let $f(x)$ be an odd function, then

$$\int_{-a}^{a} f(x)\,dx = 0.$$

(viii) Let $f(x)$ be an even (odd) function over the interval $(-a, a)$, then if the derivative exists, it is odd (even) on this interval, i.e.,

$$f(x) \text{ even} \Longrightarrow \frac{df(x)}{dx} \text{ odd},$$

$$f(x) \text{ odd} \Longrightarrow \frac{df(x)}{dx} \text{ even}.$$

A.6 ABSOLUTE VALUE FUNCTION

Let a be a nonzero real number. The absolute value of a is defined to be

$$|a| = \begin{cases} a, & \text{If } a > 0, \\ 0, & \text{If } a = 0, \\ -a, & \text{If } a < 0. \end{cases}$$

An alternative and sometimes very useful definition is to use the expression

$$|a| = \sqrt{a^2}.$$

The following properties follow directly from the definition of the absolute value and may be generalized to functions:

$$|a| \geq 0$$
$$|-a| = |a|$$
$$|ab| = |a||b|$$
$$|a^n| = |a|^n$$
$$|a + b| \leq |a| + |b|.$$

A.7 DIFFERENTIAL EQUATIONS

A.7.1 General Linear, First-Order Ordinary Differential Equation

This differential equation takes the form

$$\frac{dy}{dx} + P(x)\,y = Q(x),$$

where $P(x)$ and $Q(x)$ are given. The general solution is

$$y(x) = Ce^{-\int P(x)dx}$$
$$+ e^{-\int P(x)dx} \int e^{\int P(x)dx} Q(x)\,dx,$$

where C is an arbitrary constant.

A.7.2 Bernoulli Equations

This equation is nonlinear and can be written as

$$\frac{dy}{dx} + P(x)\,y = Q(x)\,y^n, \quad n \neq 1,$$

where $P(x)$ and $Q(x)$ are specified for a given value of n. The nonlinear transformation

$$u(x) = [y(x)]^{(1-n)},$$

reduces the Bernoulli differential equation into a linear equation of the form

$$\frac{du}{dx} + P_1(x) = Q(x),$$

and this equation can be solved using the technique given in Section G.1.

A.7.3 Riccati Equation

This first-order, nonlinear differential equation is

$$\frac{dy}{dx} = y' = q_0(x) + q_1(x) y + q_2(x) y^2.$$

where $q_0(x)$, $q_1(x)$ and $q_2(x)$ are given, and it is assumed that $q_2(x)$ is non-zero and has a first derivative. The nonlinear transformation

$$v(x) = q_2(x) y(x)$$

gives

$$v' = v^2 + R(x) v + S(x),$$

where

$$S(x) = q_0(x) q_2(x), \quad R(x) = q_1(x) + \frac{q_2'(x)}{q_2(x)}$$

If $u(x)$ is defined as

$$v(x) = -\frac{u'(x)}{u(x)} = -[\operatorname{Ln} u(x)]',$$

then $u(x)$ is a solution to the following linear, second-order differential equation

$$u'' - R(x) u' + S(x) u = 0.$$

Thus, the solution to the original Riccati differential equation is

$$y(x) = -\frac{u'(x)}{q_2(x) u(x)}$$
$$= -\left[\frac{1}{q_2(x)}\right] [\operatorname{Ln} u(x)]'$$

A.7.4 Linear, Second-Order Differential Equations with Constant Coefficients

This class of differential equations has the form

$$a_0 \frac{d^2 y}{dx^2} + a_1 \frac{dy}{dx} + a_2 y = Q(x),$$

where (a_0, a_1, a_2) are given constants and $Q(x)$ is a given function. Consider first the case where $Q(x) = 0$, i.e.,

$$a_0 y'' + a_1 y' + a_2 y = 0.$$

This equation is generally called the homogeneous part of the differential equation. The expression

$$a_0 m^2 + a_1 m + a_2 = 0,$$

is called the characteristic equation (CE) and determines the solutions to the homogeneous differential equation as follows:

(a) If the two roots to the CE, m_1 and m_2, are real and distinct, then the general solution is

$$y_H(x) = C_1 e^{m_1 x} + C_2 e^{m_2 x},$$

C_1 and C_2 are arbitrary constants, and where the corresponding homogeneous solution is written as $y_H(x)$.

(b) If the two reals are real and equal, then

$$y_H(x) = (C_1 + C_2 x) e^{mx}, \quad m_1 = m_2 = m.$$

(c) If m_1 and m_2 are complex conjugates, i.e.,

$$m_1 = m_2^* = a + b_i, \quad i = \sqrt{-1},$$

then

$$y_H(x) = \begin{cases} A e^{ax} \cos(bx + B), \\ \text{or} \\ e^{ax}[C_1 \cos(bx) + C_2 \sin(bx)], \end{cases}$$

(A, B, C_1, C_2) are real arbitrary constants.

For $Q(x) \neq 0$ and the known differential equation,

$$a_0 y'' + a_1 y' + a_2 y = Q,$$

is called an inhomogeneous, linear, second-order, constant coefficient differential equation. Its general solution may be written

$$y(x) = y_H(x) + v(x),$$

where $v(x)$ is a particular solution to the inhomogeneous equation. In general, for an arbitrary function $Q(x)$, no $V(x)$ can be found such that it may be expressed in terms of a finite combination of the elementary functions. However, if $Q(x)$ consists of linear combinations of functions whose derivatives consist of a finite set of linearly independent functions, then $v(x)$ can be calculated and expressed as a linear combination of these linearly independent functions. For this case, the following two rules may be applied to find particular solutions:

Rule 1: Let no term in $Q(x)$ be the same as in $y_H(x)$. For this case, $v(x)$ will be a linear combination of the terms in $Q(x)$ and all their independent derivatives.

Rule 2: Let $Q(x)$ contain a term that, ignoring constant coefficients, is x^k times a term $y_1(x)$ appearing in $y_H(x)$, where $k = 0, 1, 2, \ldots$. The corresponding particular solution, $v(x)$, will be a linear combination of $x^{k+1} y_1(x)$ and all its linearly independent derivatives that are not contained in $y_H(x)$.

A.7.5 Fourier Series

Let $f(x)$ be the periodic function of period $2L$, i.e.,

$$f(x + 2L) = f(x).$$

Assume that the following integrals exist

$$\int_0^{2L} f(x) \cos\left(\frac{k\pi x}{L}\right) dx, \quad \int_0^{2L} f(x) \sin\left(\frac{k\pi x}{L}\right) dx,$$

for $k = 0, 1, 2, \ldots$. The formal Fourier series of $f(x)$ on the interval, $0 < x < 2L$, is given by the expression

$$f(x) \sim \frac{a_0}{2} + \sum_{k=1}^{\infty} \left[a_k \cos\left(\frac{k\pi x}{L}\right) + b_k \sin\left(\frac{k\pi x}{L}\right) \right],$$

where

$$a_k \equiv \left(\frac{1}{L}\right) \int_0^{2L} f(x) \cos\left(\frac{k\pi x}{L}\right) dx,$$

$$b_k \equiv \left(\frac{1}{L}\right) \int_0^{2L} f(x) \sin\left(\frac{k\pi x}{L}\right) dx.$$

Definition 1 *A function $f(x)$ is said to be piecewise continuous on the interval, $a \leq x \leq b$, if this interval can be partitioned into a finite number of subintervals such that $f(x)$ is continuous in the interior of each of the subintervals and $f(x)$ has finite limits as x approaches either end point of each subinterval from its interior.*

Definition 2 *A function $f(x)$ is said to be piecewise smooth on the interval, $a \leq x \leq b$, if both $f(x)$ and $f'(x)$ are piecewise continuous on $a \leq x \leq b$.*

Theorem 1 *Let $f(x)$ be piecewise smooth on the interval, $0 < x < 2L$. Then its Fourier series is*

$$f(x) = \frac{a_0}{2} + \sum_{k=1}^{\infty} \left[a_k \cos\left(\frac{k\pi x}{L}\right) + b_k \sin\left(\frac{k\pi x}{L}\right) \right].$$

This series converges at every point x, in the interval $0 < x < 2L$, to the value

$$\frac{f(x^+) + f(x^-)}{2},$$

where $f(x^+)$ and $f(x^-)$ are, respectively, the right- and left-hand limits of f at x. If f is continuous at x, then the Fourier series of f at x converges to $f(x)$.

Bibliography

BOOKS

A. Andronov, E. Leontovich, I. Gordon, and A. Maier, *Qualitative Theory of Second-order Dynamical Systems* (Wiley, New York , 1973).

F. Brauer and J. Nohel, *The Qualitative Theory of Ordinary Differential Equations* (Dover, New York , 1989).

S. Chow and J. Hale, *Methods of Bifurcation Theory* (Springer-Verlag, New York , 1982).

B. Hassard, N. Kazarino, and Y. Wan, *Theory and Application of the Hopf Bifurcation* (Cambridge University Press, Cambridge , 1980).

J. H. Liu, *A First Course on the Qualitative Theory of Differential Equations* (Pearson Education, Upper Saddle River, NJ, 2003).

J. Marsden and M. McCracken, *The Hopf Bifurcation and Its Applications* (Springer-Verlag, New York , 1976).

R. E. Mickens, *Mathematical Methods for the Natural and Engineering Sciences*, 2nd Edition (World Scientific Publishing, Singapore , 2017).

V. Nemytskii and V. Stepanov, *Qualitative Theory of Differential Equations* (Princeton University Press, Princeton, NJ , 1960).

A. Panfilov, *Qualitative Analysis of Differential Equations* (Theoretical Biology, Utrecht University, Utrecht , 2010. https://arxiv.org/pdf/ 1803.05291

A. D. Polyanin and A. I. Zhurov, *Separation of Variables and Exact Solutions to Nonlinear PDE's* (CRC Press, Boca Raton, FL , 2022).

S. Strogatz, *Nonlinear Dynamics and Chaos* (Addison-Wesley, New York , 1994).

HANDBOOKS

H. J. Bartsch, *Handbook of Mathematical Formulas* (Elsevier Science and Technology Books, Boston, 1974). *CRC Standard Mathematical Tables and Formulas*, 33rd Edition, D. Zwilliniger, editor (CRC Press, New York , 2018).

I. S. Gradshteyen and I. M. Ryzhik, *Tables of Integrals, Series, and Products*, D. Zwillinger, editor (Elsevier, Boston , 2015).

A. Jeffrey and H. Dai, *Handbook of Mathematical Formulas and Integrals*, 4th Edition (Academic Press, New York , 2008).

G. A. Korn and T. M. Korn, *Mathematical Handbook for Scientists and Engineers* (Dover reprint edition, 2000; McGraw-Hill, New York , 1968). *NIST Handbook of Mathematical Functions*, Oliver, D. W. Lozier, R. F. Boisvert, and C. W. Clark, editors (Cambridge University Press, Cambridge , 2010).

A. D. Polyanin and A. V. Manzhirov, *Handbook of Mathematics for Engineers and Scientists* (Chapman and Hall/CRC Press, Boca Raton-London, 2006).

A. D. Polyanin and V. F. Zaitsev, *Handbook of Exact Solutions for Ordinary Differential Equations*, 2nd Edition (Chapman and Hall/CRC Press, Boca Raton, London , 2003).

D. Zwillinger, *Handbook of Differential Equations*, 3rd Edition (Academic Press, Boston , 1977).

WEBSITES

The following website provides detailed information on many types of equations appearing in the natural and engineering sciences:

Eq World, The World of Mathematical Equations.

In particular, known exact general and particular solutions are given along with the references to the source of these solutions.

Another important website is https://dlmf.nist.gov

which provides the link to the NIST Digital Library of Mathematical Functions (DLMF).

A comprehensive software system for general computational issues and many other related things is Wolfram Mathematica.

Index

1-dim, nonlinear oscillators, 206
2-dim lattice parameters, 191

A

Absolute value function, 260–261
Algebraic relations, 253–255
Amplitude, 214
Asymptotic solution, 183
Averaging method, 206

B

Bernoulli equations, 261
Bessel functions, 96, 121
Bifurcation, 137–142
　examples from physics
　　lasers, 150–151
　　statistical mechanics and neural networks, 151–152
　examples of elementary, 142–143
　　saddle node, 143–144
　　subcritical pitchfork, 148–150
　　supercritical pitchfork, 147–148
　　transcritical, 144–147
　Hopf-, 152–158
　qualitative properties, 159
Boundary-value problem, 161–162
　approximation to, 165–167
　properties of, 162–165
　qualitative methods, 167
Burgers' equation, 118–119

C

Calculus relationships, 258–259
Chebyshev polynomials, 96, 121
Coulomb damping force, see Coulomb friction force
Coulomb damping oscillator, 213–215
Coulomb friction force, 213
Cubic equations, 254–255
Culling, 222, 227–229

D

Damped cube-root oscillator, 60–62
Damped harmonic oscillator, 56–57
Derivatives and integrals, 257
Differential equations, 261–265
Differentiation, 258
Dry friction force, see Coulomb friction force
Dynamic consistency, 3–4

E

Even and odd functions, 259–260

F

Factors and expansions, 253
Finite difference discretization, 187, 237–240
Fisher's equation, 128–131
Fourier series, 264–265
Fundamental properties, 255–256

G

General linear, first-order ordinary differential equation, 261
Gompertz model, 30–31, 224

H

Harmonic oscillator, 53–56
Heat conduction
  functional equation model comments, 185–194
Heat PDE, approximate solution, 16–18
Hermite polynomials, 96, 121
Heuristic derivation, 206
Homoclinic orbit, 218
Hopf-bifurcations, 152–153
  closed integral curves, 154
  fixed-points, 154
  qualitative properties, 159
  theorem, 153–154
  two limit-cycle oscillators, 154–158
Hopf-bifurcation theorem, 153–154
Hyperbolic functions, 121, 257–258

I

Integration, 259
Interpretation, 12–13

J

Jacobi elliptic functions, 96, 121

K

Kelvin absolute scale, 196
Kink, 203
Korteweg de Vries equation, 125–128
Krylov and Bogoliubov, first approximation, 206–208

L

Laguerre polynomials, 96, 121

Lambert W-function, 121
Laplace's equation, 108
Legendre functions, 96, 121
Lewis equation, 160, 209
Linear, second-order differential equations, 262–264
Liouville–Green transformation, 78–80
Logistic equation, 29–31
Logistic–Holling–Leslie model, 224
Logistic–Ivlev model, 224
Lotka–Volterra model, 223, 225

M

Mathematical equation, 5
Mathematical modeling, 3
Mathematical models (MM), 185
Maxwell–Cattane equation, 194–196
  generalized aspects of, 205–206
  positivity and equilibrium solutions, 196–197
  space-independent solutions, 197–200
  traveling wave, 200–205
Method of separation-of-variables (MSOV/SOV), 104, 110–112
  Burgers' equation, 118–121
  first-order, nonauton ODE, 113–114
  nonlinear diffusion equation, 117–118
  solve and analyze the solution, 114–115
  special functions, 121–123
  three simplest linear wave equations, 116–117
MM, see Mathematical models
MNMCE, see Maxwell–Cattaneo equation
MSOV/SOV, see Method of separation-of-variables

N

Newton's law of cooling, 13–14

Nonlinear cubic oscillator, 57–60
Nonlinear damping oscillator, 212–213
Nonlinear oscillations, average method, 206
    Coulomb damping oscillator, 213–215
    higher-order corrections, 209
    Krylov and Bogoliubov, first approximation, 206–209
    nonlinear damping oscillator, 212–213
    quadratic terms oscillator, 215–222
    van der Pol oscillator, 209–212
Nullclines, 43–45

O

ODE, *see* Ordinary differential equation
One-dimensional dynamic system, 22
    applications, 27–38
        depends on, 33–35
            Gompertz model, 30–31
            logistic equation, 29–30
            radioactive decay, 27–29
            Spruce Budworm population model, 35–38
            tank draining, 31–33
    fixed-points, 22–24
    linear stability, 25–27
    two fixed-points, 24–25
Ordinary differential equation (ODE), *see* Partial differential equation

P

Parameter expansion method, 236
Partial differential equation (PDE), 102–104
    approximation of '1' and '0', 236–245
    linear ODE, 229–235
    symmetry-derived, 104
        analysis, 108–110
        heat conduction, 104–106
        wave, 106–108
    traveling waves, 123–124
        Burgers' equation, 124–125
        Fisher's equation, 128–131
        heat PDE, approximate solution, 131–133
        Korteweg de Vries equation, 125–128
PDE, *see* Partial differential equation
Perturbation method, 236
Physical equation, 5
Physical mathematics, 2–3
Pitchfork bifurcation
    subcritical, 148–150
    supercritical, 147–148
Poisson's equation, 108
Powers of trigonometric functions, 257
PPS, *see* Predator–prey systems
Predator–prey systems (PPS)
    culling, 227
    culling the predator, 227–229
    general properties, 224–227
    predator–prey models, 223–225
Product and sum relations, 256

Q

Quadratic equation, 253
Qualitative methods for differential equations
    calculate solutions, 10
        analysis, 18–19
        heat PDE, approximate solution, 16–18
        interpretation, 11–12
        Mickens–Newton law of cooling, 13–14
        radioactive decay, 10–11
        second-order, linear, 14–16
    dynamic consistency, 3–4
    experiments and physical measurements, 2
    mathematical modeling, 3
    mathematics and physical mathematics, 2–3
    local behavior of functions, 6

physical and mathematical equations, 4–5

R

Radioactive decay, 10–11, 27–29
  functional equation model comments, 185–194
Random walk equation, 191
Rayleigh equation, 209
Reaction–diffusion–advection, 175–176
Real physics, 191
Riccati equation, 262

S

Saddle-node bifurcation, 143–144
Schrödinger equation, 122
Second-order, linear differential equation, 14–16
Selected functions, expansions, 255
Semi-stable (SS) fixed-point, 24
Separation-of-variables, see Method of separation of-variables (MSOV/SOV)
Simple predator–prey model, 65–67
SIR model, 69–74
Spruce Budworm Population Model, 35–38
Spurious limit-cycles, 209
Stable node, 24, 46, 48, 51, 225, 226
Standard expansion () methods, 209
Sturm–Liouville problems, 76–77, 90–91
  eigenvalues and eigenfunctions properties, 91
  expansion of functions, 92–93
  first-derivative term, elimination, 77–78
  Fourier series, 94–96
  Liouville–Green transformation, 78–80
  orthogonality of eigenfunctions, 91–92
  reduction to, 93
  separation and comparison results, 85–87
  special functions, 96–97
  TISE, sketches of wavefunctions, 97–100
  vibrating string, 80–81
    both ends fixed, 81–82
    both ends free, 83–84
    one fixed and one free ends, 82–83
  vibratory modes, 84–85
Sum of angles relations, 256

T

Taylor series, 191
TFE, see Thomas–Fermi differential
Thomas–Fermi differential (TFE), 168–174
Time-independent Schrodinger equation (TISE), 87, 97–101
TISE, see Time-independent Schrodinger equation
Transcritical bifurcation, 144–147
Traveling wave solutions (TWS), 104
Trigonometric relations, 121, 255–257
Truly nonlinear oscillator, 240, 243–245
Two-dimensional dynamical systems, 41–42
  examples
    angular frequency, 62–64
    damped cube-root oscillator, 60–62
    damped harmonic oscillator, 56–57
    harmonic oscillator, 53–56
    nonlinear cubic oscillator, 57–60
    simple predator–prey model, 65–67
    SIR model, 69–74
    van der Pol equation, 67–69
  first-integral and symmetry transformations, 45
  fixed-points, 43
  linear stability analysis, 50–51
  nonlinear systems, local behavior, 51–53

nullclines, 43–45
  phase-plane construction, 49–50
  trajectories, 45–49
Two limit-cycle oscillators, 154–158
TWS, *see* Traveling wave solutions

U

Unstable node, 24, 46, 48, 51, 225, 226

V

Van der Pol equation, 67–69, 157, 209–212

Variable scaling, 177
Verhulst–Pearl model, 224

W

Wave equation, 106, 108
Wave front behavior, 177, 180–184
  approximation, 184–185
  square-root dynamics, 175
  TW solutions, 178–180
  variable scaling, 177

Printed in the United States
by Baker & Taylor Publisher Services